One Planet Is Enough

Rune Westergård

One Planet Is Enough

Tackling Climate Change and Environmental Threats through Technology

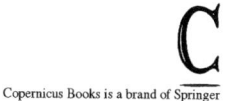

Springer

Copernicus Books is a brand of Springer

Rune Westergård
Korsholm
Finland

ISBN 978-3-319-60912-6   ISBN 978-3-319-60913-3 (eBook)
DOI 10.1007/978-3-319-60913-3

Library of Congress Control Number: 2017944301

© Springer International Publishing AG 2018

This work is subject to copyright. All rights are reserved by the Publisher, whether the whole or part of the material is concerned, specifically the rights of translation, reprinting, reuse of illustrations, recitation, broadcasting, reproduction on microfilms or in any other physical way, and transmission or information storage and retrieval, electronic adaptation, computer software, or by similar or dissimilar methodology now known or hereafter developed.
The use of general descriptive names, registered names, trademarks, service marks, etc. in this publication does not imply, even in the absence of a specific statement, that such names are exempt from the relevant protective laws and regulations and therefore free for general use.
The publisher, the authors and the editors are safe to assume that the advice and information in this book are believed to be true and accurate at the date of publication. Neither the publisher nor the authors or the editors give a warranty, express or implied, with respect to the material contained herein or for any errors or omissions that may have been made. The publisher remains neutral with regard to jurisdictional claims in published maps and institutional affiliations.

Printed on acid-free paper

This Springer imprint is published by Springer Nature
The registered company is Springer International Publishing AG
The registered company address is: Gewerbestrasse 11, 6330 Cham, Switzerland

A New Picture of the World

Imagine you could take a snapshot of the entire world. What would you see?

You would probably look upon a picture of war and misery, starvation and disaster—and the threat of irreversible climate change.

Perhaps you would have the impression that we will soon be paying dearly for exploiting more than nature can provide, and for failing to respect the limits of our planet's ecosystems.

This concept of the world is shared by a lot of people today.

However, I genuinely believe that such a worldview is distorted, because it has been taken with the wrong kind of camera. A snapshot cannot do justice to the transformation processes that have already been set in motion to make our planet a better place to live.

The threats so vividly depicted in the media today have made many people question modern civilisation, especially the concept of consumption. According to a survey made for the World Wildlife Fund (WWF) by the consumer organisation Cint, no less than 80% of young people in Sweden are deeply worried about the world's future climate. This is unfortunate as young people should have an optimistic view of the future, rather than one riddled with angst. Most people today share this bleak view of the state of the world, despite the fact that things have actually been steadily improving with each passing decade.

Positive changes can be difficult to discover with only a snapshot view of the world. To obtain a more complete picture of the progress made, our planet needs to be seen as a sequence of moving pictures instead.

Throughout history, there have been occasions when humankind has believed that it was on the brink of catastrophic disaster. Yet time and time again, development has changed the course of history in unexpected ways, and solutions to old problems have suddenly materialised.

We have all heard the question: How many worlds would be needed if everybody lived like the average citizen in the developed world?

The answer must be: One!

The Right Remedy for a Coughing Planet

This book is not an attempt to deny the difficulties. Today, some of our world's natural resources are being overexploited in ways that deserve serious attention. The whole planet is coughing, and the reason is undeniably the impact our civilisation is having on ecosystems.

The solution, however, is not to return to a primitive way of life. On the contrary, the one thing that can save us from irreversible climate disaster is new and more efficient technology. The fact that our technologically oriented society appears to create serious problems does not mean, however, that we have turned down a blind alley! As our knowledge and technology improve, we will also become able to solve the environmental problems, which is the only viable road to follow into the future. Fortunately, technology is evolving faster than ever and in a more environmental-friendly direction; an evolutionary process driven by consumption and economic growth. As such, consumption—in combination with suitable incentives or other political steering measures—is required to speed up the technological solutions needed to save our planet.

My experience from a long professional career in the field of energy and environmental technology has taught me that technology is not an enemy of the environment. Rather, technology is actually the key to overcoming the problems we are facing today.

I decided to write this book as a result of my interest in community development and a fascination for the forces behind technological advancement. Many years ago, I founded a technical consulting company called Citec, which I managed for 20 years. The company was originally started as a summer job for two people. Today, the company employs 1300 people in 11 countries.

The type of projects I worked on included the creation of designs for water purification and recycling plants, as well as the implementation of ways to generate power and fuel from waste. These projects brought me and my associates to many different countries, both rich and poor, across several continents.

Our company has also been involved in hundreds of power plant projects. In this way, I have personally witnessed the consequential improvement of people's health and welfare—all over the world—when provided with electricity.

During my professional career, I realised that it is quite possible to solve energy and environmental problems, and that industry can play a leading role when working in cooperation with well-balanced governmental control.

The explosion of knowledge and technological development that is currently in progress has convinced me that we will be able to create a balance between the planet's ecosystem and modern civilisation—and that this feat may be simpler to achieve than most of us believe.

Instead of simply registering the looming deficits in our global balance sheet, we need to remind ourselves of the tremendous progress that humankind has achieved in the past. In this way, we will be better prepared to find out how we can turn this single planet into an adequate home for our growing population.

How this can be done—and the forces available to meet these challenges—will be discussed in this book.

Korsholm, Finland
February 2017

Rune Westergård

Acknowledgements

Many thanks to my editors who have helped me to straighten out the text material and for all their generous advice and support. Specifically, I want to thank Anna Jeanne Söderlund and Lars Rosenblad at Morgan Digital, and Bert-Ola Gustafsson, Henry Nygård, Kristoffer Gunnartz, Michael Smirnoff, Berndt Schalin, the Calidris publishing company and I would like to thank collectively the team at Springer International Publishing AG and Springer Nature for support and advice in this project. I would also like to thank Gunnar Redmalm for the English translation and Paul Wilkinson for the English editing. And special thanks to my dear friends and family for their opinions and suggestions, and for their patience with my research work and writing.

Contents

1	**The Ecological Shackle**	1
	At an Important Crossroads	2
	Is Sustainable Development a Sustainable Concept?	4
2	**The Children of Technology**	7
	Technology Makes a Difference	9
	Culture Is Technology-Driven	10
	Man—A Domestic Animal	11
	The Expansion of the Technosphere Is Accelerating	12
3	**Darwin and the Machines**	15
	The Biological Evolution	17
	Technology and Biology Are Shaped by the Same Hand	19
	The Natural Consumption Selection	21
	Gene Flow and Diversity in the World of Technology	23
	The Mammoths of Mechanics	24
	How Will the Techno-Evolution Proceed?	25
	A New Species?	26
4	**The Long and Winding Road to a Better Life**	29
	From Glass to Penicillin	30
	From Stones to Apps	31
	From a Stone on a String to GPS	32
	The Pattern of Destiny	33
	A Digital Giant Leap	35
	The Phone Booth in Your Breast Pocket	35
	The Machinery of Welfare	36
5	**The Mechanisms of Progress**	39
	Let Obsolete Technology Die	42
	Maturing Technology	43

		Page
	The Sharing Economy	44
	When Products Become Services	44
	Environmental Effects Due to the Growth of the Technosphere	46
6	**Consumption—A Primordial Force**	47
	Creative Consumption	48
	Consumption Research with a Twist	49
	The Right Diagnosis for the Right Cure	50
	The Luxuries of Yesterday Become the Commodities of Today	52
	The Involuntary Luxury Consumer	54
	Active Consumption Supports Sustainable Development	55
	Accommodation, Beef & Car—The ABC of Personal Climate Consideration	57
	Buy a New Fridge and Save the Planet	59
	Change Your Car—Change Technology	60
	The Consumption of Today Creates the Improvements of Tomorrow	61
	Is Downshifting Possible?	63
	Service Consumption	64
	Growth—A Part of the Natural Order	66
	Exactly What Is Wasteful Consumption?	67
	Is Happiness for Sale?	68
7	**Real and Imagined Threats**	71
	Focus: The Problem	73
	Real and Imagined Threats	74
	Facts, Propaganda and Hidden Messages	75
	What Is the Real State of the Planet?	76
	Is the Notion of Sustainable Technology-Driven Growth Over-Optimistic?	78
	Pessimism Does not Support a Sustainable Development	79
8	**The Welfare Debt and the Rebound Effect**	81
	Does an Accelerating Development Increase Inequality?	82
	Suspension of Imports or Growth?	83
	Do the Rich Countries Have an Environmental Debt?	84
	The Rebound Effect—A Push to the Next Level	86
	Sustainability Gains Despite the Rebound Effect	89
9	**Technology Requires Freedom and Responsibility**	91
	The Carrot and the Stick	92
	Means of Control and Creative Destruction	94
	Paralysing Analyses	95
	Care and Testing	96

10	**Resources Are Dwindling—Yet Growing**	97
	The Process Box of Civilisation	98
	Infinite and Finite Resources	99
	The Copper War that Never Started	99
	From Waste to Resource	102
	Growing Resources	103
	Structural Capital and Big Data	105
	Artificial Resources Growing at High Speed	106
	The Essence of Technology—A Natural Resource	107
	Artificial Resources—A Summary	108
11	**The Climate Issue Can Be Solved**	111
	Energy Supply and Energy System Services	112
	The Research Situation Concerning the Greenhouse Effect	113
	Decoupling from Kaya	114
	Facing Challenges, Seeking Opportunities	116
	Energy Production and Technology that Reduce Climate Gases	117
	Increasing Savings Demands	121
	Super Technology	122
	The Ace up the Sleeve	123
12	**Scenarios for Success**	125
	The Politics of Climate Challenges	126
	Technology Shift Now or Later?	129
13	**One Planet Is Enough**	133
Appendix: More Future Power to Change the Planet		137
Literature		145
Sources		149

About the Author

Rune Westergård I am 64 years old. I was born and raised close to the sea in Korsholm near Vaasa in Finland, where I still live together with my wife Annika. We have two sons who are now grown-up and have left home.

I belong to the Swedo-Finnish part of the population, and my native language is Swedish. I love a life close to nature, travelling and natural science literature.

Evolution and the future of humankind are subjects that I find particularly interesting. I have always been engaged in issues regarding the future and the natural sciences and I have worked with technology for most of my life, even if I was dreaming of a career as a rock musician when I was in my teens.

My professional career as an engineer began with product development in the fields of mechanics and computerised monitoring systems. Fairly early, my creative ambitions also made me an entrepreneur and businessman. Citec, my technological consulting company, started as a summer job for two people. Today, it has 1,300 people employed in 11 countries.

Five years ago, I sold the majority of my shares in Citec, while remaining on as a board member and minority shareholder. This has given me time for research and writing.

My interest in community development and a fascination for the forces behind technological evolution have influenced me a great deal, which is the reason that I decided to write this book. With this I hope not only to make a contribution to the debate, but also to inspire some optimism and confidence in the future, especially in the younger generation.

Chapter 1
The Ecological Shackle

Abstract A concept widely used today is "the ecological footprint". This is basically the calculated space needed to provide for every individual—in terms of resources, production, consumption and waste. The weakness of this valuation of ecological footprints is that the products and services we develop and consume are not considered to generate any environmental or other specific benefits at all. This distorts the message and forces people to believe in false solutions that may well defy their intended purpose.

There are many ways to discover the size of your personal footprint; a quick internet search, for example, will provide several tools specifically for this purpose. The result is presented as "planet units", i.e. the number of planet Earths needed if every individual lived the way you live.

For my part, I have used the WWF's climate calculator to estimate my ecological footprint. Based on the values I entered, I got the result that 3.5 Earth planets would be needed if everyone in the world today lived the way I live.

That corresponds well enough with the average result for all people in the Nordic countries.

Then I made a test to see what would be needed to bring this figure down to a single planet.

I made the following assumptions regarding this hypothetical lifestyle: I am a vegan, I buy no food that is not locally produced and I don't drive a car or a motorbike. I never travel by air and I travel a maximum of two hours by train and one hour by bus each week. My family of four people is crowded into 40 m^2 of living space and our home is heated only by firewood and district heating. We buy environment-classed electricity, switch off the low energy lights and other electrical equipment when not in use and are satisfied with a room temperature of 15–18 °C.

I spend at most 300 euro a year on pets, jewellery, cosmetics, and tools, and I haven't bought a TV set, a mobile phone, a washing machine or any other technical equipment in the last year (but I have, admittedly, indulged in one single item of furniture).

I use a compost and I recycle all household waste as far as possible.

However, the WWF calculator reported that this isn't enough by far.

If we all live crowded together in cold flats and deny ourselves most of the good things that life has to offer, we would still consume an environmental equivalent of 2.1 planets.

Other studies, for example the MIPS[1] method, have come up with roughly the same results.

Looking at Finland, the average citizen's material footprint is 40 tonnes per year, and calculations made by the scientist Tuuli Hirvilammi (according to the MIPS method) show that even relatively poor people in Finland have a footprint of 18 tonnes a year.

Based on this information, we can deduce that only a homeless vegan can manage a sustainable existence, because only an individual like that can get away with an ecological footprint of less than eight tonnes a year.

So what is our conclusion?

Even if we all make our best efforts to save the planet by changing our lifestyle, actions like improved waste separation and less frequent air travel will be far from sufficient. What is needed is a fundamental revolution of our entire way of life, in a material sense.

Such an accomplishment would require far more than mere marginal modifications of behaviour, such as stopping smoking and eating red meat. The apparent solution would be to deny ourselves virtually all forms of consumption that we are used to having in our daily lives.

The biggest drawback of these calculation models, however, is that they don't take into account the negative consequences of a reduced consumption.

The weakness of this valuation of ecological footprints is that the products and services we buy are not considered to generate any specific benefits at all, either directly, indirectly or in the long run. This distorts the message and forces people to believe in false solutions that may well defy their intended purpose.

At an Important Crossroads

There are two ways to avoid increasing greenhouse gas emissions. One is to refrain from using any activity that demands energy. The other is to continue the activity, using better technology instead. Simply put, you can either stop taking hot showers, or you can heat your water with more efficient technology. In the first case, you will sacrifice some of your comfort. In the second case, you will keep your comfort, but

---

[1]The **MIPS method** (Material Input Per Unit of Service). This is the total amount of natural resources consumed by a specific product or service.

the crows sitting on your newly insulated roof will miss the heat they enjoyed in the past. The first case involves a changed lifestyle, the second doesn't.

Virtually everything we do produces emitted climate gases, because almost all current energy-dependent systems are based on energy technologies that generate such gases. In a different world, where transportation and power generation would be free from such emissions, disputes about the proper amount of car driving would be unnecessary.

Therefore, the answer is to find the solution in a redirection of technology, not in reduced comfort.

Whenever a demand for a specific product exists, the product will be further developed. Unfortunately, despite the fact that the effect of such development is beneficial to all concerned, no developmental value is granted in the calculations of an ecological footprint. The only things taken into account are drawbacks.

Let's assume that in an attempt to limit climate gas emissions, we avoid buying certain products.

One such example could be to stop buying electronic wristbands for heart rate monitoring and calorie counting. What happens then? Well, of course the industry would stop developing this product. As a result, we would not only lose the benefits of heart rate monitoring, but also of any future technologies that may appear as spin-offs from the heart rate monitor concept. Wearable sensors for the medical measurement of blood values and the revolution of remote self-diagnostic devices are already appearing on the market today, with cheaper and more efficient healthcare as an added bonus.

This branch of technology can mean a lot to our health in the future. But it depends on the simple first steps, like wristbands for heart rate monitoring.

The point is that the consumption of electronic gadgets like wristbands is actually sustainable technology in the true sense of the word, despite all the ecological calculations that seem to suggest the opposite. On the other hand, it is admittedly a development that wouldn't be possible without a certain amount of consumption of natural resources and various emissions.

A common argument when discussing ways to reduce the ecological footprint is to try to buy second-hand products instead of new. In my opinion, this is flawed logic; the best policy for sustainable global development—and a reduced footprint in the long run—is *active consumption*. As such, we should preferably buy new products with consideration of what is environmentally sound.

On the other hand, the use of second-hand products may be the only economically viable alternative, for an individual consumer. But this is an entirely different issue that has nothing to do with the ecological footprint and a sustainable global development.

As seen with the example of the heart rate monitor, the consequences of banning the consumption of certain products will have detrimental effects. Another way to see this clearly is to use a backward time perspective. Let's assume that we would have based our actions 20 years ago on the estimates of ecological footprints and similar calculation models used today. Which products would we then have to live without?

The answer is that we would have been denied a plethora of useful everyday devices such as smart phones, flat screen televisions and new energy effective machines. In fact, only a handful of very wealthy individuals would ever have had access to the electronic and data-oriented devices and services that most of us can barely imagine living without today.

The sad truth is that if the estimation of ecological footprints and similar calculation models had been allowed to control consumption, then the climate and the environment would be in an even worse state than it is today.

Virtually all products manufactured today are more environmentally-friendly than the versions that were used 20 years ago. If our civilisation today would be based on the technology level of 1995, the atmosphere would contain a much higher amount of greenhouse gases, and our natural resources would have been depleted to a much higher degree.

The conclusion is that the ecological footprint, when used as an argument against consumption, may turn out to be an ecological shackle.

Is Sustainable Development a Sustainable Concept?

The concept "sustainable development", as originally coined by the environmental scientist and writer Lester Brown, has been widely accepted by politicians worldwide. The concept was first globally acclaimed by the so-called Brundtland commission that was assigned by the United Nations in the mid-1980s. The intention was to formulate an all-encompassing view of the world's resources and environmental challenges.

In the report "Our common future", where the result of the commission's work was presented, "sustainable development" was defined as follows:

> Sustainable development seeks to meet the needs and aspirations of the present without compromising the ability to meet those of the future.

The concept of "sustainable development" is today widely used as an argument against consumption and growth. The interpreters of this concept speak of three dimensions of sustainability: ecological, economic, and social.

The meaning of the concept is that development isn't sustainable if it hurts people, destroys the environment we live in, or consumes our finite natural resources.

Some argue that ecological sustainability is the most important of the three dimensions, as it is a necessary prerequisite for both social and economic sustainability.

This message is easy to understand and accept when presented in general terms. But if you take a closer look at the meaning of ecological sustainability and welfare development, things immediately become more complicated.

According to the definition of the Brundtland commission, the meaning of ecological sustainability appears to suggest that our generation cannot make use of the planet's finite resources, because we would then limit the possibilities of later generations. According to this perspective, these natural resources should be left intact for all eternity, consequently making them utterly worthless.

However, isn't it obvious that it would be impossible to achieve social and economic sustainability without making any use of our finite natural resources? In that case, the priority of ecological sustainability becomes meaningless.

Some are of the opinion that we can actually use a reasonable part of the finite resources, as long as we don't do it to such a high degree as is the case today. But then we must answer the question of how big this reasonable part would be, and there we have no consensus.

What if we are to assume that the things created from the natural resources are actually what make the development in all three dimensions sustainable in the long run?

Minerals, metals and oil provide us with washing machines, ambulances and computers, just to mention a few concrete examples. At the same time, we indirectly create knowledge and continued technological evolution. The benefits of this may be worth more for future generations than the natural resources that were consumed in the process.

In my opinion, we can manage a sustainable evolution in the long run, even if the development in a more limited time perspective seems to be unsustainable.

But let's first have a look in the rear-view mirror to consider how humankind's development in the past has proceeded hand-in-hand with technology.

Chapter 2
The Children of Technology

Abstract It is important to keep in mind that our ancestors, before the technological era, were animals. Technology has not only contributed to the creation of *Homo sapiens*; it has also granted us the opportunity to live long and protected lives. Can a continued technological evolution create a balance between a modern way of life and the planet's ecosystem? I believe that the answer to that question is definitely yes; which is why it is of great significance to fully understand the essence of technology.

Try to imagine a group of ape-like creatures sitting huddled together somewhere on the African savannah, about three million years ago. They would have been busy using stones to crush the bones of a carcass that a pride of lions had left behind—an everyday task to access the nutritious marrow inside. On this particular day, however, something unusual happened. Instead of hitting a bone fragment, one of the creatures smashed a stone against another stone.

The other members of the group would probably have considered such behaviour as counterproductive, since there is nothing edible inside a stone. However, by breaking the stone our inventive hero managed to create a much sharper stone fragment—a hand tool—making it possible to obtain food faster and easier than all the others in the group. The stone-breaking method was soon adopted by the entire group, thus establishing the first technological innovation in human history.

This is how we imagine that technological development began, a journey that has transformed us from primitive primates[1] to space travellers. Incredibly, the catalyst that set off this long chain of development almost certainly came from a simple tool used in the immediate fight for survival.

As people today become increasingly cocooned in the technosphere[2] they are finding it increasingly difficult to step outside of their reality and view things from

---

[1] **Primates** are the group of animals that includes apes and humans. Together with certain whales, the primates are the animals with the largest brains and highest intelligence.

[2] **Technosphere**, a term for the parts of the earth's surface and adjacent areas that are affected by technological processes.

another perspective. For many, such a task is almost on a par as it would be for a fish to explore the world outside of the hydrosphere.

The technosphere is our world, our natural habitat.

It is only by studying the first technological breakthrough when hominines[3] began to make tools, that we can better understand how, and why, our material environment is continuously expanding around us.

The impact that technology has on humankind is much stronger than we tend to believe, and it touches upon the very soul of what actually makes humans human. The very idea that humanity can be regarded as a *creation* of technology, for example, is a supposition that is bound to cause great controversy—the traditional objection being that man is capable of thoughts and feelings, and is not a semi-robot. People advocating a different view to this line of reasoning are often derided as unemotional simpletons with a naive belief in science and technology—a type of reaction reminiscent of the upset Darwin caused when he first presented his theories, dethroning man from the divine pedestal.

Nevertheless, the fact that technology has played an important role in the rise of *Homo sapiens* is a fairly well-accepted thesis among scientists in the diverse fields of archaeology, anthropology and evolution. However, a source of considerable controversy between the disciplines remains when it comes to deciding on exactly *how* important that technological part is.

A common misconception in early human evolution theory was that the brain and intelligence of our ancestors were developed before tools were manufactured. Such a claim, however, has been contradicted by later evolutionary research and archaeological findings. For example, a study by Daniel E. Lieberman and Katherine D. Zink from Harvard University—recently published in Nature—supports the hypothesis that early stone tools were created long before there were hominines with brains of a size comparable with ours.

Consequently, the evolutionary step where the *Homo* genus was separated from the other primates could not have taken place without technology. (The technology we are dealing with here is defined as all artificial objects, including primitive artefacts like the simple stone tools that were in use approximately three million years ago.)

The archaeologist Timothy Taylor, professor at Bradford University in the UK, writes in his book *The Artificial Ape* that the evolution of our genus started thanks to stone tools. These tools also contributed to the development of traits like walking upright, shortened arms, less body hair, a larger brain and increased intelligence.

An ape's skull is formed the way it is because apes have to tear and chew their food using their teeth and powerful jaw muscles. Generally, apes don't prepare their food with tools as do the members of the *Homo* genus. The *Homo* genus represents the evolutionary line from the early hominines to our current species, *H. sapiens*.

---

[3]**Hominines**, ancestors of man, the evolutionary branch of primates that was separated from the chimpanzee branch several million years ago.

Earlier along the same line we find the *Homo habilis*, the *Homo erectus*, the *Homo neanderthalensis*, and others.

Technology has directed the evolution of the *Homo* genus in a very concrete way—it has affected the way we look and, above all, the size of our brain and the way it is programmed. An average human brain today is about twice as heavy as a hominine brain from 2.5 million years ago.

Chimpanzees—our closest living relatives—and other apes have not undergone the same dramatic change as *H. sapiens*, because their evolution has not been accelerated by the use of artificial tools. Thanks to stone tools, the members of the *Homo* genus became better at foraging and were able to consume more nutritious substances, like important fats and proteins, which were needed to increase the weight and volume of their brains. Current research (e.g. the study by Daniel E. Lieberman and Katherine D. Zink from Harvard University mentioned earlier) has shown that, first and foremost, it was plentiful and more nutritious food that modified the hominines' physiology and anatomy.

Technology Makes a Difference

According to Taylor and other archaeologists, two parallel evolutionary processes were probably interacting with each other in small steps. The result was a development spiral, where simple technology nurtured and stimulated the pre-human brain and innovative capability, which in turn empowered the development of more advanced technology.

In due course, the early hominines made several important innovations, like improving the stone tools with wooden handles. Thanks to the axe handle, the strike could be aimed and focused on the right spot, and the handle also served as a lever, augmenting the strength of the arm. The handle made a difference, and in a historical perspective it was an innovation of the same or even greater importance than any advanced machine today.

Consequently, this interaction between biological and technological evolution is a proven driving force behind the evolution of the species *H. sapiens*, which appeared on the stage about 200,000 years ago. Without technology, we wouldn't have looked the way we do, and we wouldn't have been the most intelligent species on earth. In fact, we wouldn't even exist. This may seem strange, but without all the artificial objects that we associate with technology, the evolution would have followed entirely different tracks. As it were, humankind has created itself with the use of technology.

Another important milestone in humankind's past was the gradual conversion from a life as hunter-gatherers to an agricultural society. It is plausible that men first became domiciled and started to cultivate the earth sometime between ten and twenty thousand years ago. The original hunter-gatherers were nomads and their nativity was low, but when they became settled, their populations skyrocketed. In additional to the larger food supply, the growing number of people also affected evolution, because the genetic variation increased.

Once again, technology was a prerequisite for development, this time in the form of all sorts of picks, spades, and levers. In the area of the Middle East known as the Fertile Crescent, archaeologists have found harvesting tools made of honed volcanic glass, that remain sharp even according to modern standards. These innovations, tools and stationary cultivation—in combination with a beneficial climate—were the conditions required for a lasting agricultural society. The agricultural revolution was one of the most important steps in the rise of civilisation.

We represent the first species of this planet that—to a large degree—has evolved by artificial means. Our development is still ongoing, not least because of medical technology.

Culture Is Technology-Driven

It is not only human biology that has been affected by technology; our entire culture—including lifestyle and behaviour—has also been affected. This is contrary to the traditional way of describing human development, where artificial objects are considered to be products of a culture, rather than the opposite.

Tools, clothing, vessels and fundamental building principles have kept our ancestors warm, fed and reasonably healthy, while simultaneously driving the development of culture in various directions. In this light, it is possible to state that virtually all major technical innovations have guided the development of culture into new paths.

It is easy to imagine how the simple artefacts that made it possible for the early hominines to control fire half a million years ago created new ways of social interaction. Sitting around a fire created a feeling of security, which undoubtedly brought along a new sense of togetherness and solidarity. Some scientists have suggested that this was also the period when language development rapidly advanced. It is an interesting thought. Perhaps a distant genetic memory of such a time may explain why we still experience a special kinship when sitting around a camp fire together. Who knows—perhaps even the first sparks of romance were ignited around those early fires!

Research has shown that at the time we are dealing with here, members of the *Homo* genus began to create more sophisticated tools, and of a greater variety than before. Other things that appeared at about the same time were body-painting, jewellery and other items of beauty—early portents of the cosmetics and fashion concepts of our times.

Our ancestors gradually became more social-minded and more ready to cooperate, even with different tribes. Tolerance increased and aggression decreased.

Findings of early baby carriers and similar artefacts show the extent to which various *Homo* species at various times have been able to create aids and tools to facilitate everyday life, all the way back to the beginning three million years ago.

The baby carrier was particularly important since it freed women's hands for other tasks, which made various food-related work easier.

Thanks to the baby carrier, which is considerably older than the 200,000-year-old *H. sapiens*, the early hominids could have children whose heads continued to grow after birth. The children could be born in an unfinished condition, so to speak, meaning that the head would achieve its full size outside of the womb.

If the head grew too big in the womb, the child (and probably the mother) would die. The child would also die if the birth was premature, before the skull was fully developed. This is where the baby carrier appeared on the scene making it possible for the "prematurely" born babies to survive.

This is yet another example of a technological aid that guided the *Homo* genus' evolution on the way towards eventually becoming beings with larger brains—*H. sapiens*.

Man—A Domestic Animal

Given the culture and lifestyle of today, man is totally unable to survive without the help of technological aids. As a species, man is like a domestic animal, utterly helpless if placed alone in the wilderness. This is actually not a new idea at all, since there is one thing that we can be certain of: *H. sapiens* have *never* lived in a kind of harmony with nature. On the contrary, we have always been in a constant state of war against nature.

It may seem odd that nature should be considered humankind's adversary, but we have to remember that the concept of nature includes a lot of things, not all of which are pleasant: predators, insufficient food, bacteria, sickness and bad weather. In humankind's persistent struggle against nature, we have prevailed thanks to our talent for innovation and technological aids. Nevertheless, it is import never to lose sight of the fact that nature is essential for our survival.

Our thinking capability has developed through an intimate interaction between ourselves and the artificial world with its artefacts that we have created around us. In this respect, our thinking does not exist separately from our artificial surroundings.

The philosophy professor Andy Clark at the University of Edinburgh argues that the *H. sapiens* species' cognitive development and talent for problem solving are products of technology, rather than the opposite.

This becomes more understandable when we consider the development of a baby's brain. Small children who are stimulated by interaction with their parents and play with physical objects are simultaneously programming their brains. Conversely, children who for some reason have been deserted or isolated, suffer from a lack of stimulation; they develop fewer brain cells and synapses and are far more liable to behavioural disorder.

The making and use of tools by those early hominids obviously stimulated the evolutionary development of their brains in a similar way. A recent discovery to support this theory was made when a group of scientists tried to persuade some apes to create tools by knocking stones together. They found that the stone-knocking activity stimulated exactly the parts of the brain where the understanding of language is situated, which supports the theory that language and toolmaking share a common evolutionary background. The fact that apes lack this ability to understand language is something that may explain the great difference between apes and men.

This integrated interaction between man and his artificial world is obvious when we study the Renaissance. The great improvements in this period—i.e. scientific progress, increased urbanisation and travel, a virtual explosion of information interchange thanks to the invention of the printing press etc.—occurred simultaneously with a revolution in art and literature.

Today, culture is changing at a faster pace than ever before. It would be a huge understatement to merely claim that the culture development of today is empowered by the internet, computers and social media. A more comprehensive description would be that today's culture is actually *created* to a very high degree by modern technology, if not exclusively so. Billions of people spend a large part of their time in interaction with social networks. In fact, nothing in the earlier history of mankind can be compared with the cultural revolution currently in progress.

The Expansion of the Technosphere Is Accelerating

The pace of technology is accelerating, for better or worse. However, although technology is the source of our anthropogenic environmental problems, it can also provide us with the solutions.

Today, we have greater possibilities than ever before to control the development of our species and culture, through technology in general and gene technology in particular. Unfortunately, there is no certain way of knowing where the technology era—which started with a few simple stone tools three million years ago—will bring us in the long run. However, a retreat to a simplistic existence with minimal technology seems to be an extremely improbable outcome.

One of the world's leading genealogists, Craig Venter, suggests that humankind is currently entering a new phase of evolution. Other scientists argue that humans today have assumed control over their own evolution—or that we are playing God, as the critical voices claim. Gene technology may seem frightening, but it also creates enormous possibilities for progress. And in reality, there is nothing new with the idea that we can affect our own evolution by artificial means. As mentioned earlier, we already started doing this at least three million years ago.

It is important to keep in mind that our ancestors, before the technological era, were animals. Technology has not only contributed to the creation of *H. sapiens;* it has also granted us the opportunity to live long and protected lives.

We have now arrived at a point in time when this development is being questioned. The modern way of life—with high levels of consumption—has made the planet cough, which scares some people out of their wits.

Many believe that development is no longer sustainable, due to the ecological footprints we are creating. However, in all probability, the forces of evolution that have provided the good things in modern life can also eliminate the drawbacks!

Can a continued technological evolution create a balance between a modern way of life and the planet's ecosystem? I believe that the answer to that question is definitely yes; which is why it is of great significance to fully understand the essence of technology.

Chapter 3
Darwin and the Machines

Abstract The fact that technology is subject to evolution is undisputed. Thus, a brief summary of techno-evolution would be that it all started with hand tools. Then the machines came, which were followed by the intelligent machines we have to some extent today. These intelligent machines, in turn, will one day be superseded by the autonomous machines anticipated in the near future. The most fascinating thing about this techno-evolution is the driving forces behind it.

Life has been found in every reach of our planet—from the bowels of the earth to the clouds in the sky. It is incredible to imagine that the earth was once totally barren, until life somehow emerged from inert matter. This process was started by a force which has empowered the development of life to an amazing degree ever since—the same force that has modified tiny micro-organisms and transformed them into space travellers. A force that has coloured the planet green and filled every drop of the oceans' water with life.

What is the force being described?

The force of evolution, of course.

Technology is also subject to evolution. And even here, we can observe a "survival of the fittest". Technology evolution (abbreviated as the "techno-evolution") is a process that occurs gradually over a long period of time and is—to a large degree—autonomous.

You can say that technology is modified in three dimensions—evolution, innovation and product development. See Fig. 3.1.

- *Innovation* represents a breakthrough, something new—not merely an idea but rather a realised idea. Thus, an innovation is an implemented invention, which has occupied a position in the technosphere. As such, a technological innovation can be regarded as a leap in development, although it can still be traced from earlier technology.

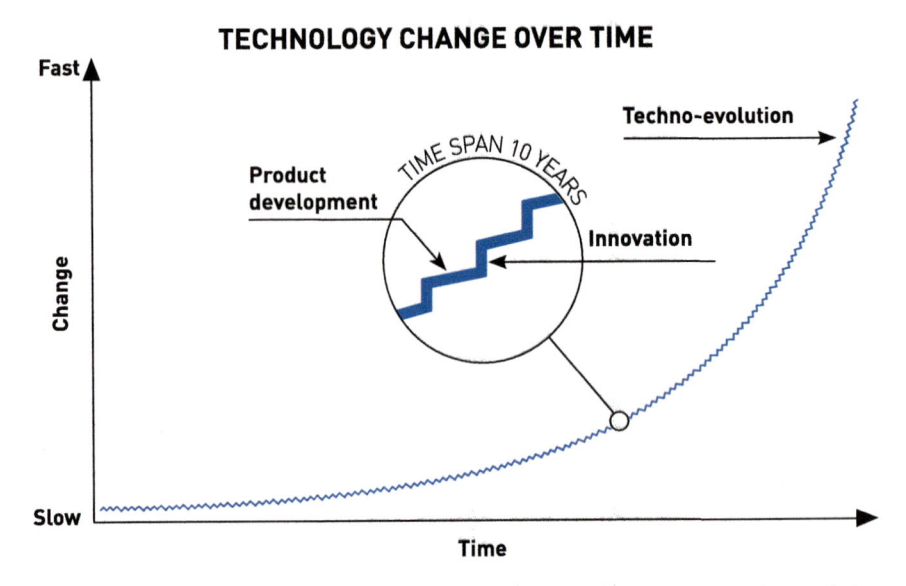

Fig. 3.1 Technology is changed by innovation, product development, and techno-evolution. While innovation and product development are concrete changes in a short time perspective, the techno-evolution is a process that proceeds gradually over a very long time span and without intentional human interference

- *Product development* is different in the sense that it can continue over a long period of time without being an innovation. Small ongoing improvements of existing products are not designated as innovations, although such a continuous improvement process may still be able to raise a product to new levels of excellence over time.

The biggest difference between these dimensions is that techno-evolution isn't as obvious as the other two; it is more difficult to observe and most of us never even notice it. The fact that it is self-guiding and occurs over a long period of time is sometimes confusing. Furthermore, it is a distorting mirror that may easily provide a false image of the world we live in.

A still image simply doesn't work. You can't capture a process in a single moment.

Techno-evolution will create as many revolutionary changes on this planet as biological evolution has ever done.

The impact on the planet is already clearly visible—on the one hand we see negative effects on the environment, while on the other hand, it is possible to observe unbelievable progress for humankind.

The techno-evolution began three million years ago, when our ancestors created the first stone tools mentioned earlier. This evolutionary primordial force is not destructive in itself, as we might tend to believe when witnessing climate and environmental problems in the world. If that had been the case, the *Homo sapiens* species would have become extinct a long time ago.

3 Darwin and the Machines

Without exaggeration, we can conclude that the future survival of humankind will be decided by technology, as we are totally dependent on technological systems and aids of various kinds. Water purification, electricity, medical equipment, and communication utilities are just a few examples of technology that has become indispensable today.

And that isn't all. Actually, the future of the entire planet will be decided by the specific technology that is developed and implemented in the current century. We have even been forced to ask ourselves if we have now reached a stage in development where the technological consumption society has run aground.

With regard to that, it is strange that the techno-evolution concept never seems to appear in debates concerning the climate and environmental situation. Personally, I am convinced that techno-evolution is the key to achieving a harmonic relationship between civilisation and nature on our planet.

So what actually is techno-evolution?

To provide a satisfying answer to that question and to understand how this concept affects our civilisation and planet, we have to realise how the technological evolution process actually works. To make this easier, we can compare it with biological evolution.

The Biological Evolution

The generally accepted theory is that biological evolution began on earth a little more than 3.5 billion years ago, with single molecules being combined in various ways. This reaction was started by environmental changes due to volcanic eruptions, meteorites and other natural disasters.

The molecules were chemically and physically combined into increasingly complex groups. Somewhere in this process, one member in the enormously varied mass of molecules became self-reproductive—it started to change itself so that a copy was created. For obvious reasons, these self-reproductive (but still not living) molecules became more frequent. As they were not perfect, some copies were slightly corrupted. As a result, new variants continuously appeared, and these variants were sometimes even better at reproducing themselves than the original had been. The complexity increased continuously, and after hundreds of millions of years, something resembling life was created.

When Charles Darwin published his book *The Origin of Species* in 1859, he introduced a theory that explained how life on earth had been developed from its primitive origins to the overwhelming diversity that exists today.

Biological evolution can be described in a simplified way as a process where the organisms that are best suited to the environment in which they live manage to prevail and multiply more efficiently than other organisms. This natural selection is the key factor in the evolution process that results in a gradual modification of all biological life.

Even if it may seem strange, we have the changes of environment and climate to thank for our existence. If earth had been an absolutely stable living environment, the evolution would have followed an entirely different path, and complex organisms like mammals and men probably wouldn't have existed today.

About 65 million years ago, earth was hit by a large meteorite—a disaster that caused a dramatic change of climate. As a result, the dinosaurs became extinct and mammals became free to evolve. Without this climate change, the dinosaurs would probably still be the undisputed masters of earth. This was neither the first nor the last time that climate changes created by natural phenomena led to the mass extinction of some species.

This shows us that climatic changes should be taken very seriously, because they always cause enormous changes, for better or worse. One species may come out a winner, another a loser. Unpleasant and even chaotic changes are sometimes necessary to fuel the evolutionary creation process.

Technology is also subject to the same phenomenon. The techno-evolution has been at work for at least three million years—slowly at first, accelerating later. Scientists and writers who have discussed this subject, for example John Ziman and W. Brian Arthur, have shown that there are significant similarities between biological and technological evolution.

Evolution is a complex process, involving several sub-processes. While some scientists believe that the biological and technological evolution processes closely resemble each other, others point out the differences that actually exist.

In my opinion, when we look at the fundamental principles, the evolution processes in biology and technology are more similar than dissimilar. Especially in a longer time perspective, much seems to indicate that the basic principles are fully universal.

The renowned biologist Richard Dawkins has shown that culture is also subject to evolutionary processes. Dawkins suggests that culture is changed by evolution when exposed to variation and selection.

As an example, different languages and dialects have been created by evolution. When a person begins to pronounce a word differently than before or invents a new word, this is analogous to a mutation. If other individuals follow the lead and reproduce the new pronunciation or word, it will spread throughout the population. Even if the linguistic differences at first seem insignificant, the cumulative changes will be considerable after one thousand years. This explains why we in the northern parts of Europe can no longer understand the old Nordic languages that were spoken by our ancestors, the Vikings.

But what is the basic reason behind the technological evolution? Which forces are driving it forward?

Will technology submit to our control, or is it really an independent power?

Technology and Biology Are Shaped by the Same Hand

The fact that technology is subject to evolution is undisputed. Therefore, the metaphor of a development tree—something often used to illustrate biological evolution—is also relevant for technology. The trunk is simple but new branches appear with time. One thing leads to another, all the way up (Fig. 3.2).

In light of the fact that most common sub-processes in biological evolution theory have their counterparts in technology, the following comparisons between these two areas of science have been made to help to comprehend how new technology develops:

- *Intensification*—How complex organs, for example the eye, could originate from evolution has puzzled science for a long period of time, and has sometimes been used as an argument against evolution theory. Today we know that the eye developed in several small steps through so-called intensification. It began with a mutation that provided a primitive organism with a singular light-sensitive cell, which gave some kind of advantage in the natural selection process.

The offspring of this organism inherited this light-sensitivity, and because of the advantage, this variant became more numerous than its competitors. Over time, a

BIOLOGICAL EVOLUTION	TECHNOLOGICAL EVOLUTION
Modern human being	Satellite
Human-like ape	Aircraft
Ape	Car
Mammal	Bicycle
Reptile	Wooden cart wheel
Fish	Stone wheel
Multicellular life	Stone axe
Unicellular life	Shaped stone
Natural material	Natural material

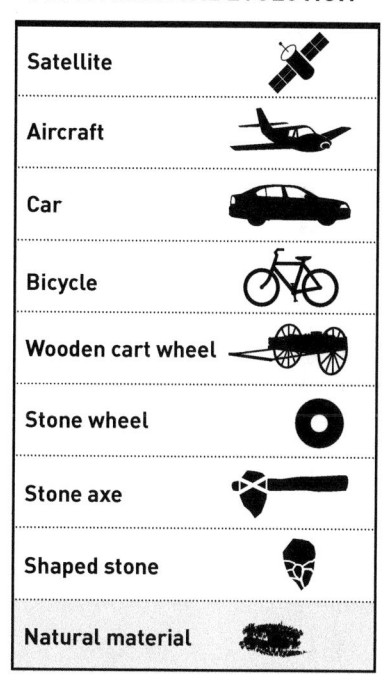

Fig. 3.2 Technological products and biologic life both began with non-living natural material. The evolutionary processes that control their development are similar, in that one step leads to another, with ever increasing complexity

kind of light-sensitive spot appeared, and additional small changes resulted in muscles that could move the spot. Thus, there were lots of intermediate stages between the first light-sensitive cell and the eye.

Such intermediate stages of various organs are not mere assumptions, but easy to prove, as they can still be observed in simpler organisms today. A classic example that has been proved using fossils are fish fins, which later developed into legs, and further into bird wings.

There are several counterparts to this in the field of technology, and one of the best examples is the computer. If we trace the computer's development far enough back, we realise that the intensification started with a collection of small stone beads in a wooden frame. In between, there have been countless intermediate stages on the way to the modern computer.

The concept of intensification also helps us to understand the difference between development and evolution. The word "development" describes concrete short-term changes, while "evolution" is used for slow, gradual changes. It is these gradual changes that we sometimes find hard to observe and understand.

- *Existence environment*—All life on earth exists in the so-called biosphere. This place is also known as the "life environment" or "existence environment", and is subject to constant change at the whim of nature – this includes everything from daily weather variations to disasters such as impacts from giant meteorites. All environmental changes drive evolution forwards as organisms have to adapt to new environments for survival.

Likewise, technology has its own existence environment.

This environment is affected by many different factors, including man's lifestyle and knowledge, the structure of society and the climate.

- *Reproduction*—The *genes* are the basic building blocks in the *genome* that transfer the characteristics from one generation to the next. The reproduction process itself is generally controlled by DNA, a long molecule chain of genes that could be described as a set of instructions or a recipe.

Technology's equivalent to a gene would be the part of a specific technology that is reproduced. A new innovation is generally not 100% new. Rather, a new piece of technology is usually born via a chain of development, and is founded on older predecessors or an earlier known technology, as is shown in Fig. 3.3.

Richard Dawkins has studied analogues between biological and cultural evolution, and has introduced the *meme* concept as a non-biological equivalent to a gene. A "meme" is the smallest unit of cultural transfer, a unit of imitation. Consequently, the basic technological solutions that are transferred and copied could well be called "techno-memes".

- *Mutation*—When a cell splits, reproduction errors sometimes occur. This change is called a mutation. Without mutation there would be no evolution, as mutation is the only source for new variants of living organisms. Mutations either occur spontaneously or are triggered by disturbances such as virus attacks, radioactive radiation or foreign substances that have entered the organism.

Technology and Biology Are Shaped by the Same Hand 21

Fig. 3.3 The genes are the entities which transfer properties to the next generation. The technological version of a gene could be defined as the part of a technology that is copied to later versions. A new technological product is born via development in several steps from earlier technology

The technological equivalents to mutations are ideas. Let's have a look at a common example: the light bulb. It has been told that the inventor Tomas Edison tried 1000 different ideas for a working light source before finally choosing an idea that fulfilled the criteria he had stipulated after development and testing. The type of lightbulb that proved to be successful represented the actual innovation, while the remaining 999 experiments were discarded variants, like the mutations that don't survive natural selection.

In the field of biology, those mutations that don't lead anywhere are considerably more common than the ones spawning a persisting change. Nature is generous by allowing an overwhelming number of experiments, but extremely picky when it comes to salvaging the results. And exactly the same is true for product ideas—most are scrapped. A technological idea that is never realised is similar to a mutation that doesn't result in a successful change of the organism.

The Natural Consumption Selection

When a mutation occurs, we get several similar variants of the same organism. They are launched into a competitive situation, and the variant that is best adapted to the living environment wins. The new improved organism will spread and transfer the successful properties to new generations. The old organisms lose the competition and disappear. Darwin chose to call this process natural selection, to separate it from the choices breeders make to improve the quality of livestock.

In the context of techno-evolution, natural selection has an equivalent in *consumption selection*, which includes everything from simple household products to advanced technological systems. The selection on the market—representing the sum of many consumers' actions—is a phenomenon that is as important for techno-evolution as natural selection is for the evolution of organisms. Exactly like biological evolution, technology is adapted over time to its new existence environment as part of a continuous process.

New products are exposed to enormous hazards when passing the various stages on the way from an idea to an established position on the market. Most won't survive the journey. A typical inventor is a creative person with a head bursting with ideas. Most of the ideas created are discarded by the inventor himself, but some are passed on. A first selection among several "mutations" has then already been made in the inventor's brain. The following selection is often made in a product development situation, where several brains are involved. This is the time when a lot of research, laboratory work and testing goes on.

Finally, we arrive at the most important level of selection—the consumers' verdict. The products that most successfully satisfy the consumers demands will survive, while others eventually become obsolete, i.e. "extinct". A product has to be well adapted to its market (its existence environment) to be chosen.

This process very closely resembles the mutations and natural selection in biological evolution.

As mentioned above, the concept of consumption works like natural selection in the biological world and is totally decisive for evolution. The choices consumers make affect how a product develops over time. Smartphones, for example, have developed incredibly quickly because hundreds of millions have been sold each year. Naturally, the forces of market selection will affect so-called consumer products faster and more profoundly than, for instance, medical robots. In cases like that, it will be a team of specialists who make the selection, even though they also represent a kind of market.

Natural selection can establish some organisms as long-term survivors, while others rapidly change, depending on the strength of the selection pressure. A current example is resistant bacteria. The penicillin group of antibiotics—one of the most important medicines of all time—has saved millions of people from sickness and death. Consequently, the use of such antibiotics has increased enormously around the world, while resistant bacteria have simultaneously multiplied at an alarming rate. Through the extensive use of antibiotics, the bacteria had been subject to an increased selection pressure—their environment had literally turned into a death trap. But even in this deadly environment there existed a few strains of bacteria with mutated genes that made it possible for them to survive the antibiotics. After a while, these germs multiplied and further mutated through the selection pressure to become even more adapted to an antibiotic environment. The result today is that some antibiotics are no longer effective.

In stable environments, on the other hand, organisms hardly change at all—a state of equilibrium is established. Mutations will simply not create any advantages in an environment without any changes or selection pressure. If we apply this situation to techno-evolution, we can observe that products and technology develop and modernise at a pace in keeping with an increasing market demand—i.e. when the products are consumed at a faster rate.

The opposite is also true—development slows down if the selection process is not working, for example when consumption decreases. Technology may even be frozen for long periods when consumption is poor.

Gene Flow and Diversity in the World of Technology

Throughout history, we can observe that very different kinds of boats were developed in places where there was no contact with other civilisations. For example, the *sampan* was invented in ancient China—a boat that was unknown in Europe—while the *knarr* of the Vikings was unheard of in Asia. These vessels were improved and developed in separate directions over hundreds of years.

In this day and age—when such isolation has been breached—designs are copied and spread in an almost contagious way, so that boats all over the world are becoming increasingly similar in many respects. Still, in an apparent paradox, the room for new variations of boat properties and looks is also increasing globally.

In other words, a certain product may be manufactured in lots of variants co-existing all over the world, even if the traditional local types also survive locally for a considerable period of time.

This can be compared with the biological concept of *gene flow*, which means that genes are transferred between different populations of the same species. The result is that the differences between populations decrease, while the common gene pool becomes larger and more varied.

Technology that is developed in an isolated environment without external influences is somewhat reminiscent of biological inbreeding.

It is easy to see how increased travel and trade during the last millennium has affected technology. Useful technological solutions were copied and spread like wildfire after lying dormant for hundreds or even thousands of years. Today, the risk of technical inbreeding is negligible.

Biological diversity—or *biodiversity*—is a popular concept that is considered to be important and worth preserving by virtually everyone. However, no one seems eager to defend the phenomenon of technological diversity with similar fervour. Still, techno-diversity may be just as important for humankind as biodiversity—one doesn't have to exclude the other.

For products to be improved over time, there is not only a need for market demand but also for diversity. Diversity means that several competing variants of the same product type are released on the market. Each one of these products may have a certain advantage over the others, and the consumers' selection will decide their destinies.

Therefore, the seemingly endless supermarket shelves filled with lots of similar items serve a purpose, so we shouldn't frown upon such growing diversity.

If we are enchanted by the multitude of flowers cropping up every springtime, then in all honesty, we should also embrace the wide selection of barbecue grills in the hardware stores with the same admiration. The diversity of spring is always a joy to behold!

The Mammoths of Mechanics

No less than 95% of all the biological species that have ever existed on earth are now extinct, e.g. the sabre-toothed tiger, the mammoth and the dinosaurs. In most cases, this has been caused by natural processes.

The rise of human civilisation is another factor that is known to affect biological diversity on the planet—organisms failing to adapt to the consequential changes of their life environment due to human activity will disappear, exactly as evolution theory describes.

A third reason is that invading competitive species take over.

Exactly the same phenomenon exists in techno-evolution—machines that can't be adapted to changes will not make it in the long run. We all know examples of outdated equipment consigned to the technological scrapheap: typewriters, mechanical calculating machines, tele printers, telegraphs, slide rules, mechanical telephone switchboards, phonographs, VHS-recorders and carbide lamps.

A lot of machines become "extinct" simply because they are defeated by the competition.

For example, typewriter manufacturers struggled for ages to save their products, trying to improve them in various ways. But that was of course hopeless, because the new "species" of computers, armed with word processing software, proved to be a technology that was much better at transforming thoughts into printed words.

In the same way that organisms become extinct when they don't fit into their present life environment, outdated technological solutions disappear when they can no longer cope with the competition.

A particular interesting and currently "extinct" machine caught my interest in my neighbouring village Hankmo outside Vaasa in Finland. It was a mechanical machine for digging ditches called an Autodigger. It was designed in 1912 by Isak Wahlstedt and was patented in Finland, Sweden and Germany. The Autodigger was manufactured 100 years ago in Isak's home village, and was driven by a 16 horsepower petrol engine from the Wickström engine factory in Vaasa.

Although the Autodigger was a challenger for conventional ditch ploughs and manual labour, it was never a successful product. The price was high, the design was complicated, and it didn't work in stony ground. The province of Ostrobothnia —where Hankmo is located—is probably one of the stoniest places on earth; not exactly the best possible environment in which to introduce the Autodigger. It is stone dead today! Nevertheless, the Autodigger is a beautiful example of a mechanical mammoth. Figure 3.4, *a photograph of a restored Autodigger at its birthplace in Hankmo village.*

Fig. 3.4 A photograph of a restored Autodigger at its birthplace in Hankmo village, Korsholm, Finland

Still another item that has been phased out is the horse-drawn plough, despite the fact that it was an extremely environmentally-friendly product. The horse is fuelled by renewable energy and emits no harmful substances, except for a small amount of methane gas. The horse plough, and especially the horse itself, was driven out of competition due to its low work capacity; albeit its DNA has survived in modern tractor ploughs.

The thing that dinosaurs and typewriters have in common is that they are big, slow and extinct. Naturally, "extinct" is not an enviable state to arrive at, although it is indispensable in the context of evolution.

How Will the Techno-Evolution Proceed?

As we have seen, an important milestone in the history of evolution was when inanimate matter spawned life that was able to reproduce. And one of the differences that are often mentioned when comparing biological and technological evolution is, of course, that ability to reproduce.

Contrary to inanimate objects like technology, living things are able to mate or split and reproduce themselves. Until quite recently, that is how things appeared when seen in the rear-view mirror. However, such a perspective is no longer true, especially when we consider that today there are examples of self-reproducing technology, such as data viruses.

In this context, it is of interest to note that even biological evolution can trace its origins to inanimate matter. The earth was completely lifeless for 500 million years until dead molecules started the evolution process that created the basic components of future living cells. In such a light, it would be misleading to limit the concept of evolution to a process that only includes living matter.

Today, we are witnessing the beginning of a new phase of evolutionary development that may become the epoch of a new age. This is because technology that is able to reproduce itself is becoming increasingly intelligent and independent, while humans, simultaneously, are in the process of mastering biological evolution —as a developer of organisms, including the human organism itself.

The evolution process is increasingly changing from natural to artificial selection. In other words, biological processes like mutations and natural selection will not decide the development of the human species to the same degree as before. Nevertheless, these artificial interventions will also be subject to the basic principles of evolution, like the techno-evolution.

A current prediction—inspiring both interest and fear—is that machines will one day become more intelligent than their human creators. Today, we have no way of knowing the extent to which this will be realised: what we can be reasonably certain about, though, is that machines will become much better than humans at developing new machines.

In an issue of the science magazine *Nature*, published in 2015, we could read about yet another example of the progress of artificial intelligence (AI). A software algorithm had *taught itself* how to play several video games better than any human.

According to one of the world's leading experts in the field, the American inventor and futurist Ray Kurtzweil, the ultimate AI explosion—when technology will be able to improve its own intelligence more efficiently without human interference—is predicted to occur about the year 2045.

When this explosion does take place (also known as the technological singularity), the result will be a gigantic historical leap—an extremely fast techno-evolution—of cosmic proportions.

A New Species?

It is not improbable that biological and artificial life will merge in the future. There may be robots containing living matter, and even today, it is not uncommon to replace some failing body parts with artificial variants.

Today's implants, titan joints and surgically inserted electro-mechanical devices are only the beginning. We are currently well underway with a form of development that will make it harder than ever before to see the difference between biological and technological matter. Only a few years ago, such technology was regarded as pure science fiction.

Thus, a brief summary of technological evolution would be that it all started with hand tools. Then the machines came, which were eventually followed by the intelligent machines we have today. These intelligent machines, in turn, will one day be superseded by the autonomous machines anticipated in the near future.

We humans are not merely descended from the human-like apes known as hominids. We have been transformed into humans thanks to an evolutionary process integrated with technology—and there is really no way back!

When we fully accept this insight, we have every possibility to utilise techno-evolution as a tremendous resource to further improve our civilisation within the planet's boundaries.

After all, we have techno-evolution to thank for being transformed from cavemen to space travellers, with an expected lifespan that has increased from 30 to 80 years.

Admittedly, there will always be a certain element of risk involved in new technology. Lately, leading physicists and IT specialists the world over have warned that artificial intelligence can be dangerous and that it may even threaten humankind's survival.

Some have envisioned self-guided robots and missiles beyond human control. A doomsday scenario like that cannot be totally excluded, although I prefer to believe it would be rather improbable.

It may seem cynical to point out, but there might actually be some advantages with advanced technology in the arms industry, as long as it is used to reduce civilian losses by eliminating human error and improving weapon precision.

However, in all probability, the world community will have to agree on the terms for the control and limitations of AI (especially AI-related weaponry) in line with current international agreements concerning nuclear weapons. After all, weapons of all kinds—in the hands of the wrong people—are the greatest hazard of all.

In my opinion, there will be certain types of people in the foreseeable future who are far more dangerous than robots. However, the direction of AI development appears to point towards technology protecting people from themselves. Isaac Asimov once said that "knowledge itself should be used as a barrier against the dangers it brings", which is still very much relevant today.

A knife can be used as a murder weapon, but in the hands of a surgeon, it can also save lives. Everything depends on how the technology is used.

Personally, I believe that artificial intelligence will soon turn into the most useful tool in the history of humankind—a super tool that will aid us on the way to a civilisation in harmony with the planet.

If machines can become faster workers and smarter problem-solvers than men, it is an offer that we cannot afford to refuse. It is a resource that we really need, given that this single planet will soon be populated by eleven billion people.

We have already taken the first steps on that road through gene technology. We may even consider the possibility of a new species, a hybrid of an artificially designed man and advanced technology in the same body. These are speculations that might be realised in the far future, hundreds or thousands of years from now.

But even that would be a short span of time compared to the three million years that have gone since our early ancestors split stones to create the first tools.

Even if humans can affect the biological development of their own and other organisms, we are still governed by evolution in a way that cannot be fully controlled. Where this process will take us is impossible to know, although it seems improbable that modern man will be the last species in the long development chain of the *Homo* genus.

In the short term—i.e. the next 20 years and the rest of this century—we will experience an accelerated development of gene technology. This is a controversial subject due to the inherent risks involved, albeit there are also tremendous possibilities. Using gene therapy, we may be able to prevent or treat otherwise incurable diseases. If gene therapy can help us to cure cancer and other consumptive diseases, it is something that we can hardly deny ourselves. But if we use this technology to manipulate our genome in the pursuit of superior talents and abilities, we are venturing into dangerous territory with new and unknown hazards.

Most scientists find it natural to approach such challenges with humility, but some limits will still have to be agreed upon. In a longer time perspective, such limits will automatically expand, in pace with increased knowledge and further technological development.

The use of genetically modified organisms (GMO's) in agriculture can also be perilous if transgenic organisms and genes spread to natural ecosystems.

Several scientists claim that GMO's are basically no different from selective plant breeding practices that have already occurred for centuries. Still, resistance against GMO's remains strong, especially in Europe. However, for the sake of argument, let's assume that we had banned all forms of plant breeding, agricultural mechanisation, insecticides and artificial fertilisers some decades ago. Then the "Green Revolution", which started in the 1960s and saved more than one billion people from starvation, would never have happened.

GMO's are potentially able to prevent food shortages and improve the future environment. From this point of view, it may seem wrong to categorically ban such technology.

On the other hand, whether we really want to use all the other possibilities that gene technology offers—for example in doubling the human lifespan—is another matter entirely.

The heart of the matter is that we can expect new technology and knowledge in general to develop more rapidly during the next century than it has up to now, including all cumulative progress during the entire history of humankind. A development of which we are a part.

H. sapiens will probably control their own development in small steps, becoming increasingly artificial, until a possible new species, *Homo futura*, appears.

Chapter 4
The Long and Winding Road to a Better Life

Abstract The road to inventions, innovations and progress is often winding and full of surprises. In the case of antibiotics, the breakthrough was delayed because the luxury consumption of certain products was too slow. This may sound strange but there is a certain logic behind it. The list of technological innovations that have contributed to welfare and social equality could be extended infinitely. Without the printing press, for example, there would have been no general access to information and hence no democracy. During the techno-evolution process, machines and new technology have eliminated many heavy, hazardous, unhealthy, and monotonous tasks.

And now for some good news. The World Health Organization (WHO) recently published a report showing that the human lifespan has significantly increased in all parts of the world in just 10 years: in Africa by 8% and in Europe by no less than 9.7%. In the year 1900, the average lifespan globally was 32 years; in 2013 it was no less than 69 years.

The forecast for 2050 is 76 years.

Even so, there are people who believe that things were better in the past—despite the difficulty of finding any evidence to support such a claim. In fact, if you look at any past century or decade, you will have to admit that it was definitely worse before. For example, during the years 1695–1697, one third of Finland's population died of starvation because of the climate, which at that time was still totally unaffected by industrialism (even if inefficient administration was partly to blame). It was one of the worst European disasters ever; it was even more devastating than the infamous Irish famine in the 19th century in terms of the percentage of the population affected.

In the middle of the 19th century, London—with its population of 2.5 million citizens—was literally drowning in excrement. The entire River Thames had become one enormous reeking sewer. Because the outlets in the city couldn't swallow the faeces of so many people, rainwater mixed with excrement poured into cellars—as people vainly tried to protect themselves against the abominable stench with handkerchiefs or perfume. Thousands died of cholera and similar diseases due to the polluted water—only half of all new-born children born at the time lived to the age of five.

Dr. John Snow was the first to realise the connection between sickness and bad drinking water. Due to his discovery, a new sewage system was constructed and put into service in 1859.

This changed everything. The same sewage system is still working today, even though the population has now grown three times since it was built.

During the years 1918–1920, a staggering total of 50–100 million people perished in the so-called Spanish flu pandemic. On this occasion, most of the victims were fairly young, 20–40 years old. The mortality rate from the catastrophe could have been considerably reduced, however, if only vaccines and antibiotics had been invented earlier.

But why hadn't they been invented?

Well, the answer to that question is rather interesting and is as follows: It was impossible to develop vaccines and antibiotics until the basic technology of spectacles had developed to a certain level.

From Glass to Penicillin

The road to inventions, innovations and progress is often winding and full of surprises. In the case of antibiotics, the breakthrough was delayed because the luxury consumption of certain products was too slow. This may sound strange but there is a certain logic behind it.

If spectacles had been transformed into a mass market product sooner—instead of being a luxury item reserved exclusively for the rich—the microscope would have been invented earlier. This is an example of how consumption often works as an accelerator for medical and technological development.

Glass was discovered around 5000 B.C. in the Middle East. After a couple of millennia, people in Mesopotamia discovered how to shape objects of glass and make the material more transparent, which meant that it could be used for bowls and objects of beauty. The next milestone was when the Romans (circa 200 B.C.) developed the technique to polish glass and invent a new line of products, among them small glass windows. We believe that the Chinese discovered the lens around 1000 A.D., thus inventing the magnifying glass. And in the year 1280, the Italians finally had the idea of mounting two magnifying glasses into a wooden frame that could be worn on the nose—and spectacles literally saw the light of the day.

The lens was also the origin for another important innovation, the microscope, which first appeared in the late 17th century, thus enabling a breakthrough in medical science. With the microscope it became possible to study bacteria, a line of science that finally led to the British scientist Alexander Fleming's discovery of a germ-killing mould in 1928, and ultimately to the development of the medicine penicillin, introduced in 1941.

The result was a medical revolution that has saved uncountable lives since.

However, no less than 250 years passed between the technological and the medical breakthroughs. What is even more astounding, is the fact that Fleming's discovery was actually made by chance. He had forgotten to tidy up his bacteria cultures before leaving for a holiday, which gave the mould time to grow. When Fleming returned and studied his samples under the microscope, he noted that the mould had killed all the bacteria around it. Sometimes technology isn't enough to achieve progress—you need some luck too.

So what can we learn from the history of glass?

First, the path that techno-evolution follows is full of surprises and is rather unpredictable.

Second, development is accelerated by early and luxury consumption.

Third, techno-evolution tends to work in favour of material consumption being gradually replaced by service consumption. Examples of such development include the use of internet resources for meetings, banking and gaming, etc. This is very important in the context of finite natural resources and effects on the environment.

Let's now take a look at a few other concrete examples of how material products have been replaced by service solutions.

From Stones to Apps

Stones actually have a lot in common with application software for computers (apps). The first permanent form of information interchange appeared when early hominids used sharpened stones to scratch and etch the first picture-like signs on cliffs and cave-walls more than 40,000 years ago. The mobile variants, stone and clay tablets, were invented much later. With a giant leap in time to the year 2600 B.C., we arrive at the point where papyrus entered the stage; a light and flexible material enabling the use of more convenient writing tools than stones and the like.

In the Middle Ages, a material made from animal hides—known as parchment—was commonly used.

Paper, which was first manufactured in China more than 2000 years ago, slowly spread to the Middle East and Europe. Originally it was handmade, until a paper machine was invented in France in the early 19th century. Since then, paper-making has been developed and refined by mechanical and chemical processes.

From paper to newspapers, magazines and books, the journey of technological development has continued towards a digital format. Today, various e-readers, tablets and touch screens are replacing paper media at an accelerating rate. The transition from clay tablets to paper took about 3000 years, while the change from paper media to digital solutions conceivably may be over in 30 years. This change is already showing in a decreasing demand for paper.

Compared to paper, digital technology saves natural resources and reduces CO_2 emissions in both production and distribution. My guess is that newspapers and books made of paper will cease to exist as mass-market products during the coming decades.

To transfer information between people, we have thus gone from stone and clay tablets via animal hides and paper to electrons. The difference in energy and resource consumption is enormous. It is actually so staggering that it would be meaningless to try to express in figures. You may get an idea of the magnitude of this revolution if you imagine that all the information that is searched for, transferred and read on the internet each day was distributed around the world in paper form. It would, of course, be impossible.

From a Stone on a String to GPS

Whenever the Airbus A 380 (the largest passenger plane in the world) lands in Sydney—15 h after a scheduled take-off from Dallas—it will have carried almost 500 passengers to their destination.

The flight will often have been made in darkness and clouds; a feat made possible thanks to navigation technology.

The first known navigation instrument was a stone on a string that was let down from a boat until it reached the bottom. The stone was then pulled up at regular intervals so that the seafarers could estimate the distance to the shore by studying the sediments that had stuck to the stone. This method was used in Egypt in 2500 B.C. A later invention was the wooden sundial that the Vikings, among others, used to navigate the seas from about 800 A.D. The Vikings also used so-called sunstones—transparent crystals that refracted the light in a certain way, making it possible to determine the sun's position even in cloudy weather.

The magnetic needle, the origin of the compass, was discovered by the Chinese about 2000 years ago. At first, however, the Chinese didn't understand why the needle moved, and the phenomenon was only used for divination. As no one was able to explain the needle's behaviour, the Catholic Church was also suspicious of the invention after it reached Europe.

During the Middle Ages, sea captains could even be persecuted for witchcraft if they used a compass. Fear and superstition have often delayed the implementation of new technology. We could ask ourselves which useful technologies are delayed today because of fear.

The next big step of the intensification process was the sextant, used for measuring the altitude of the sun; it revolutionised navigation in the beginning of the 18th century. The following revolution in this field was the radar in the 1930s, using reflected radio waves to determine the proximity and position of objects.

In our time, the American Global Positioning System (GPS) has become the most widespread technology for navigation, based on satellites which send signals to receivers on earth. The development from a stone on a string to the GPS has proceeded in an evolutionary way, with a few revolutionary milestones and many minute modification steps in-between.

The Pattern of Destiny

At the end of the 1790 s, French soldiers brought home silken handkerchiefs and scarfs with beautiful patterns from Egypt. These exotic souvenirs were much sought-after by the French ladies and so a fashion trend was started. As demand rapidly grew, handkerchiefs began to be manufactured locally using manual looms. But then—as with virtually all technology—increasing demand and consumption created an evolutionary leap. In this case, the result was an innovation which led to even more tremendous changes than anyone could have possibly imagined. It was a way to automate the looms, using paper tapes with tiny holes to control the patterns.

In time, the punched paper tapes were also recognised as being useful for storing and transferring other forms of information. The engineer Herman Hollerith was inspired by this invention that made it possible to manage large amounts of data. He used it to design punched cards for a mechanical calculation machine that was used for the great American population census in 1890.

It was estimated that the census would have required a lot of staff and taken 10 years if managed manually, but thanks to the machine, it was finished in just six weeks.

Later, the IT company IBM developed bar code technology, which was founded on the same basic principle as the punched cards. Today, we have devices that can read significant amounts of information from stamp-sized QR codes (Quick Response codes), with their patterns of minute black squares.

The loom example clearly shows that there is a form of genetic code, a DNA you might say, that unites old and new technology in the same field. The smallest common unit of code that the loom shares with the computer is the control instruction, which is defined by programming according to the on/off principle. A paper tape or punched card contains no other information than units of "hole" or "no hole", and the interpretation of this information decides what the pattern will be.

The very first computer was also programmed according to the hole/no hole principle, implemented by perforated paper tapes or cards.

Soon after, a principle was developed so that electrical components, like vacuum tubes and later transistors, could be used for better performance. The "hole/no hole" condition was replaced by "current/no current". By assigning the value 1 to "current" and the value 0 to "no current", it was possible to design a mathematical programming language based on the binary numeral system. In this system, all values are represented by the digits 1 and 0 (current/no current) in various combinations, which is the basic principle for all computers.

It is fascinating to contemplate the complexity of biological designs like the eye, the ear, the nervous system and the brain, all marvellous creations of the natural evolution process. As mentioned earlier, the development of the eye probably started with an organism that was created, through mutation, from a single light-sensitive cell. After a myriad of small developmental steps, where light-sensitivity proved to be advantageous, the eye achieved its present form. Similar intensification steps can also be observed in technological evolution.

The noun "computer" is of course derived from the verb "compute", i.e. calculate. In the beginning, the computer was just one alternative to the mechanical devices that were used for calculations. But the computer as an idea can be traced much further back in time. The art of calculation probably started with something as simple as placing small stones in a certain pattern, or simply counting on the fingers. In fact, the words "digit" and "digital" are derived from the Latin word "digitus", finger.

The first primitive calculator was invented around 5000 years ago—the *abacus*, or counting frame, with movable beads made of stone or wood arranged on wooden sticks. The abacus became a surprisingly efficient tool for increasingly advanced calculations, as the device was further developed during several millennia.

The abacus was eventually replaced by the mechanical calculator that was developed in the 17th century. These devices were improved with components from other technologies, like cogs and gears from clockworks. This is an example of the gene flow of technology, where the corresponding biological process allows for influences between different species. Another example of gene flow between different technologies is when the punched cards from the looms were transferred to mechanical calculators in the 19th century.

The electro-mechanical calculation machine that saw the light of the day in the 1940 s was an intermediate product, a hybrid device with both mechanical and electronical elements. A few years later, the first pure electronic calculation machines were conceived and were named "computers".

Old technology does not immediately become "extinct" when new innovations are introduced; there is almost always a certain period of overlap. Consequently, mechanical calculators were still in use in the 1960 s, because the first computers were expensive and prone to operation failure.

The first "real" computer, ENIAC, inherited the fundamental concept of using punched cards for programming. Otherwise, it was based on vacuum tube technology; its weight was about equal to a fully loaded articulated lorry and its length was equal to two such lorries. Still, its capacity was not greater than that of a modern pocket calculator. The price of the machine was 500 million dollars and its energy demand was 130 kW. That power would be enough for several thousand computers today.

Since ENIAC was completed in 1946, the development of the computer has proceeded according to the so-called Moore's Law—i.e. the number of transistors on a single chip doubling every 24 months. Calculation power and memory capacity are roughly still increasing at that same rate today. The three generations of the computer's development—vacuum tubes, transistors and integrated circuits—have been evolutional leaps, where each generation has always been a necessary prerequisite for the generation following.

I have always been amazed by how often a seemingly insignificant technical invention suddenly proves to be tremendously useful in a totally different context than it was originally intended. That alone should be reason enough for having a positive view of technological diversity. Another thing that amazes me is the difficulty of trying to predict the winding future road of techno-evolution. A minute innovation today can turn out to be the embryo of a revolution tomorrow.

A Digital Giant Leap

Today, computers are vital components in virtually everything that moves, from cars to toothbrushes. They come in all shapes and sizes, manufactured by many competing companies. The diversity is large and necessary, as the society is the computers' life environment and the customers' selections decide which variants are best adapted for survival.

The rapid turnover of electronic products is what empowers techno-evolution, even if many people are frustrated by telephones and computers that become obsolete in only a few years.

The social media that have appeared in the wake of the IT revolution are a good example of how technology and culture are developed in a kind of co-evolution. A memorable fact is that the first email message was sent in 1971, another that Facebook was introduced as late as 2004.

Since then, the development of social media has proceeded at a breath-taking speed, with thousands of applications and channels and billions of new users.

Sometimes technology makes a giant leap in less than a single human lifetime.

As another example, I am reminded of the fact that my entire collection of recorded music can be stored digitally on a USB-stick weighing 10 g. In my childhood in the 1950 s, the same amount of music would have weighed four tonnes, equal to three private cars. It would have required a lorry to transport all of it!

The development of digital storage is not only appreciated by man—the planet sings along too!

The Phone Booth in Your Breast Pocket

The world record for the fastest distribution of a new technology should probably be awarded to the mobile phone. The mobile phone is a good example of how technological improvements can increase a product's performance while simultaneously reducing the need for natural resources.

Around 20 years ago, 5 kg of natural resources were consumed in order to manufacture a single telephone; today 50 g are sufficient. In other words, with up-to-date technology, the same amount of natural resources is sufficient for 50 times as many telephones. The active consumption of mobile phones in the developed countries during the first 10–15 years when the product was available on the market dramatically reduced the required amount of raw material. The material needed to provide each one of the seven billion people living on earth today with a telephone would only have been sufficient for 2% of the population 20 years ago.

During its developmental period so far, the mobile phone has become roughly 100 times lighter, smaller and less energy demanding. Simultaneously, it has become 1000 times faster, and has actually also become much cheaper. And finally, the functionality of smartphones has undergone such an exceptional development that it would be meaningless to express their usefulness in figures.

Can we conclude then that this development is unexceptionally beneficial, despite the exploitation of rare metals and resulting waste? The short answer, surprisingly, is no.

When seen in the current perspective, this development does not support ecological sustainability. However, the mobile phones of today are just an intermediate stage on the way to the even better variants of tomorrow. So to avoid stalling the development, we should continue to buy new telephones to ensure total sustainability in the long run.

The Machinery of Welfare

The list of technological innovations that have contributed to welfare and social equality could be extended infinitely. Without the printing press, for example, there would have been no general access to information and hence no democracy.

Another interesting example is the washing machine. In the past, the manual washing of clothes generally demanded one day of work per week. After the introduction of the washing machine, one hour per week was usually sufficient. As household work was traditionally reserved for women, the machines were a definite improvement—enabling women to have more time to pursue other interests.

One of the first user-friendly electrical washing machines was introduced as early as 1910. However, washing machines were still uncommon in the Nordic countries until the mid-1950s. The Cambridge economist Ha-Joon Chang and others have argued that the washing machine belongs to those inventions that have promoted the liberation of women and made it possible for them to get an education and work away from home.

The washing machine and the dishwasher are both examples of technical innovations that have changed society by furthering equality between the sexes. At the same time, hygiene has been improved and water has been saved.

However, although the machines have liberated us from many heavy tasks—both in industry and the home—the birth of the machine age was by no means painless.

Electricity was one of the greatest scientific discoveries of all time. However, it was seen as something to be feared when it first made an appearance in upper-class homes. Indoor warning signs, like the one below, were common:

The Machinery of Welfare

> This room is furnished with Electrical light. Please do not attempt to light it with a match. Just press the button next to the door.

Benjamin Harrison, the 23rd US president (1889–93), always relied on servants to turn the light on and off because of his fear of getting hurt. For the same reason, a lot of people in general avoided electric door bells for a long time.

New technology has often been associated with superstition and fear, and it is not uncommon that it meets hard resistance. Many people—especially those fearing they would lose their jobs to machines—have regarded technology as a serious threat. Ever since the first appearance of the printing press, craftsmen and workers—such as the Luddites—have sabotaged printing presses, mechanical looms and industrial sewing machines, in an attempt to stop development.

But old methods, workplaces and professions will always disappear and be replaced by the new. This is inevitable when modern and more efficient technology is created.

However, the general public is not the only sector that has shown a suspicious attitude towards technological advances. Even educated people and experts—who should have been able to see opportunities rather than problems—have occasionally uttered comments that now seem comical in retrospect. There are countless examples of inventions that are considered indispensable today—ranging from computers to planes—that were originally rejected by headstrong sceptics who should have known better. They were simply stuck in the view of the existing technology of their time. Below are a few examples:

Thomas Treshold, train engineer, 1835:

> It is highly improbable that any vehicle transporting passengers would be able to move at a higher speed than 15 km/h.

Rutherford B. Hayes, American president, 1875:

> The telephone is a great invention, but who would ever want to use one?

Rex Lambert, journalist on The Listener, 1936:

> Television will not matter in your lifetime or mine.

Naturally, all attempts to discredit or banish a useful innovation are doomed to failure in the long run. But the benefits that new technology can add to man's possibilities may temporarily work against the interests of some.

During the techno-evolution process, machines and new technology have eliminated many heavy, hazardous, unhealthy, and monotonous tasks. Our typical work today, sedentary and stressful, is not without its problems, but few would be willing to return to the heavy and strenuous physical labour of the past.

Technological development is simply a necessary prerequisite for the welfare society as we know it today.

Chapter 5
The Mechanisms of Progress

Abstract Techno-evolution will inevitably lead to higher functionality and a reduced consumption of energy and materials per item produced. In general, it is rather difficult to find much evidence to support the notion that a higher technological level causes significant local environmental problems. Without intending to trivialise incidents, I nevertheless remain convinced that technology solves more problems than it causes. A nation's transition from a primitive agricultural society based on self-sufficiency into the first stage of industrialism can lead to environmental disturbances in the country. However, such problems can be eliminated in pace with the progressing development and modernisation of the nation.

It is easy to forget that until fairly recently, humanity was constantly repressed by predators and disease carriers. Viruses, bacteria, poisonous fungi and vermin—combined with the fury of the elements—were all decisive factors when it came to whether or not the harvest would succeed and there would be food on the table.

Diseases like malaria have claimed many millions of lives throughout history, and in some cases still do. Even today, malaria alone kills approximately one million people each year. It is spread by a mosquito that thrives in swamps, which explains why the sickness in older times was known as "swamp fever". Far into the 20th century, malaria was a common cause of death in European marshlands, for instance in Holland and Spain. Since then, the ditching of lowland forests and marshlands has been an important weapon in the struggle against malaria.

We are transforming our planet for a reason! And the reason is to assume control over our lives, and to avoid being exposed to the whims of nature.

The most important reason for our longevity today is reduced infant mortality. It is difficult to imagine the despair that parents in the 19th century must have experienced when almost half their children died before the age of 15. Losing a child has always been every parent's worst nightmare.

Since then, infant mortality has been reduced, not only in the rich countries but all over the world. But there are of course still poor countries where infant mortality is unacceptably high, despite the very positive global trend.

In countries with a low and medium income, infant mortality was reduced by no less than 52% between the years 1970 and 2000. According to the research team behind the book, *How much have Global Problems Cost the World*, there are four main reasons for this:

- Higher income   3%
- Better education for women 17%
- More doctors   12%
- Technological progress   68%

Many people may find it surprising to learn that technological progress is responsible for the main part of this positive development. Until recently, techno-evolution has been serving humankind in a way that hasn't been fully recognised. Despite some limited cases of over-consumption and waste of resources, the general development continues in the direction of higher product performance and less energy consumption.

To counteract the dystopias often mentioned in the media, let's now take a brief look at a few of the many positive technological improvements that have taken place throughout history. We have already noted that the computer is one such example.

Modern lighting devices only require a small percentage of the energy that old wire filament lightbulbs needed to produce the same amount of light, Fig. 5.1. The American economist William Nordhaus has calculated that the price we consumers have to pay for interior lighting today is one thousand times lower than the price we paid to use whale-oil lamps and candles.

Refrigerators and freezers have also been dramatically improved during the last 10–15 years. As a result, energy consumption has been reduced by more than half during this short time period.

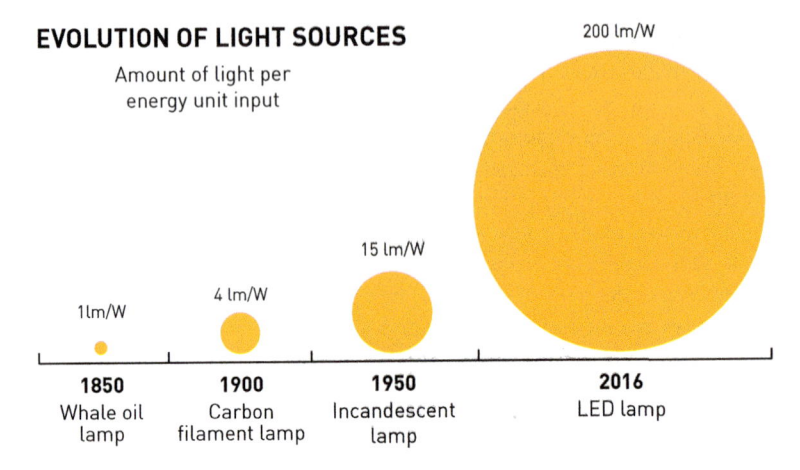

Fig. 5.1 The whale oil lamp wasted 99.5 percent of its energy in comparison with the LED-light. The technological development saved the sperm whales from extinction

Over the years, engines have been made ever lighter in terms of their weight to power ratio. And in terms of fuel efficiency, the diesel engines of today require only 1 kg of fuel to do the same amount of work that a traditional steam engine would have required 67 kg of comparable fuel to do.

This rapid technological development includes all modes of transport. The fuel consumption of planes has been reduced by 70% in the last 40 years, in terms of litres per passenger and flight distance. Train performance has also been tremendously improved since the first steam locomotives in the early 1800s. The electrical trains of today are based on magnet technology, so-called Maglev-trains. This method allows for speeds of 500 km/h or more, while simultaneously saving maintenance costs, reducing noise and offering great benefits from an environmental point of view.

The ships of today are also much more fuel efficient than those of yesteryear, thanks to a series of new technological innovations—for example the bulbous bow, a protruding formation on a ship's bow which allows the vessel to cut through the water more smoothly, thus reducing fuel consumption.

Exhaust gas emissions from cars were a big and controversial issue in the 1970s, but then the discussion was not so much focused on CO_2 emissions as on nitrogen oxides, hydrocarbons, carbon monoxide, lead and particles. At first, many argued for a general limitation of car driving, but eventually the legal regulation of exhaust emissions together with new technology came out as the solution. Catalytic converters for cars—developed in Sweden incidentally—were the most important component in this technological shift.

Right now we are experiencing another technological leap, where fossil-fuelled cars are gradually being replaced by electric vehicles. In fact, we are witnessing a new era in the evolution of the car. I believe that the days of cars running on petrol and diesel are numbered, while cars using bio fuel with an efficient power-to-fuel ratio will probably linger for some time into the future. Eventually, electric cars driven by solar-produced electricity or fuel cells will take over.

This is by no means a wild guess, as we know that most cars in the future will be bought in high-populated countries with lots of sunshine. Additionally, an electric motor has a much higher efficiency than a combustion engine, i.e. it salvages more of the input energy. The information company Bloomberg's prediction from 2016 is that in 2040, a total of 35% of all cars sold will be electric.

Of course, such a transition will take several decades. Qualified observers predict that the hybrid car—which is equipped with a traditional combustion engine as well as an electric motor—will be the harbinger of the electric car and that it will temporarily dominate the market in large parts of the world.

On a global perspective, the fuel demand of private cars and other lighter vehicles is assumed to considerably decrease during the next 25 years, given that the number of cars will continue to grow at the current rate. According to Exxon Mobile, the average consumption of all cars in operation by the year 2040 will be 44% lower per kilometre than in 2014. This means that despite the expected increase (80%) in the number of cars in the world, the total amount of fuel required will start to decrease from the year 2020. Even if the transition to electric cars takes some time, improved conventional and biogas car technology will also contribute to reduced emissions.

This is a tremendously positive development, because the actions required to make car energy consumption more effective and limit—or even eliminate—the exhaust problem in the long run are among the most important steps we can take to counter the climate threat.

It is undeniable that many technological innovations made during the industrial era have transgressed some of the planet's ecological boundaries. But in order to obtain a balanced view of the possibilities for development, we must also remember that the technological society has solved more problems than it has created.

Let Obsolete Technology Die

During my early career I used to work with diesel engines. I learned then that a big diesel engine is a product that can continue to work for many decades if the worn-out parts are replaced.

Many people engaged in the environmental debate argue that we should stop scrapping old equipment and build things to last much longer instead; the logic being that by refraining from manufacturing new things, we would be saving the earth's resources and energy.

Is this really true?

Considering the amount that diesel emissions have actually decreased because most of the world's ship-builders have renewed their fleets with modern engines, then the answer must be a resounding no.

Reconditioning old engines instead of installing new ones would not have been to the planet's benefit.

Admittedly, it is true that building new engines has meant that some metals have had to be mined from the earth; however, there is no lack of such metals and the amount mined is relatively small, especially when considering that the recycling rate of iron and steel is very high.

However, the fuel oil saved thanks to newer, energy-efficient engines being used has compensated for the energy and metals spent on the production of new equipment many times over.

A ship engine of 15 MW, constructed according to the latest technology, saves roughly 180,000 tonnes of oil during 30 years of operation compared to a reconditioned engine from the early 1980s. At the same time, the CO_2 emissions are reduced by half a million tonnes.

The natural resources (metal and energy) used to manufacture a new engine is negligible in comparison with the environmental gain. This principle is the same for virtually all energy-consuming technology.

Maturing Technology

Techno-evolution will inevitably lead to higher functionality and a reduced consumption of energy and materials per item produced. I have analysed a wide range of technology concepts from ancient times to the modern era, everything from infrastructure elements to simple consumer products, and found no evidence to the contrary. In fact, it is difficult to find any exceptions to this rule—this includes both the production process and the actual use of products and service. Furthermore, this increase in functionality and saved resources occurs without any regulations or political means of control for this purpose.

In Fig. 5.2 we can observe how evolution affects material consumption. The examples show the weight of some products today, as a percentage of their original weight. The dashed lines show a plausible future weight reduction, based on new materials, optimised structures and new production methods being developed today.

(The car example shows the weight relative to the power.)

Economists are used to expressing the same principle from their perspective, for instance that a profit-optimising behaviour by a company results in minimising the initial resources. The manufacturers who can produce a fridge with the least amount of raw material will simultaneously increase their competitive power.

All the material that isn't required for the fridge is added to the manufacturer's profit. Less raw material per produced unit also means less waste.

Additionally, development today is moving towards an increased recycling of material and energy. This was not evident during the first half of the 20th century for several reasons; raw materials were cheap, global competition was weak, knowledge was less developed and the technology level was lower.

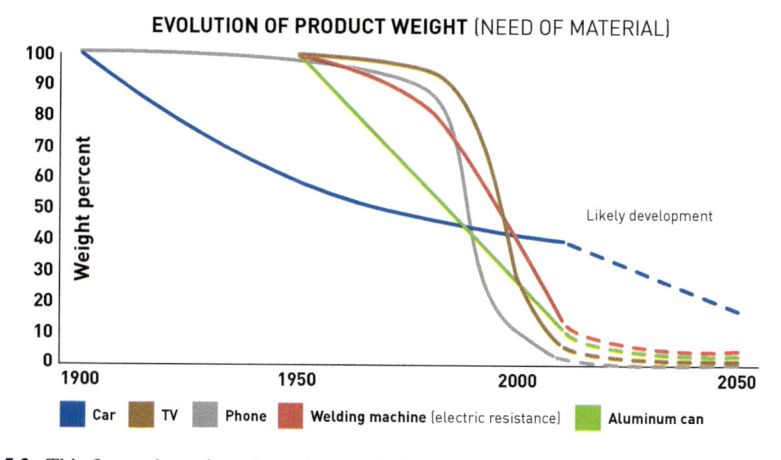

Fig. 5.2 This figure shows how the techno-evolution reduces the required amount of raw material for various products. The reduction of weight and the examples are no exceptions but are valid for most material products

As with many other changes, the beginning is slow, but the speed will accelerate with time. We will see faster changes and increased sustainability in the near future.

The Sharing Economy

The phenomenon that people exchange or sell their services and lend or borrow things via various platforms on the internet is rapidly growing thanks to modern technology. Carpooling, ride-sharing, flat-sharing, and various lending services are all examples of this growing business.

Another term for this concept is common consumption.

Sharing your expenses with others is economically advantageous for the individual consumer and also a way to use resources more efficiently. Public transport, centralised power production, the internet etc. can also be called sharing economies, albeit the actual term is specifically associated with new ways to share resources and expenses.

In this context, we may ask ourselves whether we work against development by re-using old things. As stated earlier, this is usually the case. However, the most important point here is that the individuals are allowed to rule over their own consumption as explained below.

The sharing economy, where people or organisations share and lease things, has always existed. The difficulty arises when political means of control is used to support a specific development. We may, for example, contemplate what might have happened if the political authorities in the 1980s had decided upon a massive extension of the telephone booth system, arguing that the sharing economy would be better for the planet's ecology than to allow all individuals to have a private phone in their pockets.

The point I wish to make is that sharing products and services is a good thing when the market is in control, preferably with as little political interference as possible.

When Products Become Services

The shift in the direction of less resources per functional unit is obvious. In some cases—as a result of this universal principle—some technology that was originally created as material products has been transformed into services—so-called dematerialisation.

Newspapers, books and virtually all kinds of information found online are good examples of this kind of development. Let's try to imagine an alternate reality scenario where billions of people today consume all their information printed on paper. We would be forced to witness a totally devastating deforestation for the next 20 years, despite increasingly successful paper recycling processes.

Another example deals with travelling. As we all know, travelling requires both physical vehicles and energy. Nevertheless, an increasing number of journeys are now being replaced by virtual meetings online. This is a fast growing business trend, where time, expense, and environmental considerations are the main driving forces.

During my own career, I have experienced several new technical milestones that have reduced travelling. For example, around the year 2000 it became possible for technical design teams from around the world to create constructions and drawings together, even if the engineers themselves were physically located in different countries. A team in Finland can work in real time with their colleagues in Sweden or India. The CAD system makes it possible for everyone to observe what the others are doing on screen, because they all work in the same computerised model.

Other examples of dematerialisation include the fact that toys, games, and money are increasingly being transformed into services.

If I may allow myself a speculative look into the future of dematerialisation, it is not implausible to suggest that an increasing number of musical instruments will one day be manufactured without, or nearly without, any raw materials at all. A drum kit may be replaced by several small motion detectors that are able to register hand movements in the air and transform them into sound electronically. The air guitar may soon be here for real.

The Swedish national economist Marian Radetzki has discussed dematerialisation in his books *Råvarumarknaden* ("The Market of Raw Materials") and *Människorna, naturresurserna och biosfären* ("Man, Natural Resources, and the Biosphere"). He has estimated the value of a number of typical products and services in dollars per kg. These figures are a simplified way to measure these products' contribution to the BNP. Below are some examples of price in dollars per kg (based on the values in the year of 2000):

– Petroleum	0.15
– Raw steel	0.2
– Paper for newspapers	0.4
– Private car	15
– TV set	60
– Submarine	100
– Large passenger plane	600
– Mobile computer	1000
– Mobile phone	2000
– Jet fighter	6000
– Telecommunications satellite	40,000
– Bank services	virtually limitless.

It is well known that raw materials and simple products have a lower value per weight ratio than processed products. Radetzki notes that countries that are growing richer become less dependent on products with a low price per kg, while commodities based on human knowledge, which don't weigh as much, take over an increasing share of their economy.

Environmental Effects Due to the Growth of the Technosphere

In general, it is rather difficult to find much evidence to support the notion that a higher technological level causes significant local environmental problems—even when acknowledging accidents like the Fukushima Daiichi nuclear disaster or the Deepwater Horizon oil spill. Without intending to trivialise such incidents, I nevertheless remain convinced that technology solves more problems than it causes. Nevertheless, a nation's transition from a primitive agricultural society based on self-sufficiency into the first stage of industrialism can lead to environmental disturbances in the country. However, such problems can be eliminated in pace with the progressing development and modernisation of the nation.

From the earliest of times, humans have always struggled to achieve the best possible life for themselves. Problems are solved and things are constantly improved. This is a built-in human trait, one of our basic needs, and a strong instinct.

As such, we are not helpless when hit by tsunamis, pandemics and agricultural disasters—instead, we are able to face such challenges head-on and solve them using our technology and expertise. In the midst of all the climate and environmental alarm bells ringing today, we need to remind ourselves more than ever before about the tremendous progress we have accomplished in the past.

But the fact is that the effects of our civilisation's interference are now becoming so profound that the concept of a new geological epoch has been introduced—*Anthropocene*, the epoch when man himself appears like a giant force of nature, reshaping and transforming the planet, including the functions of the ecosystems.

Transforming the planet is good for many reasons, motivated mostly by a strong desire to assume control over our lives and to avoid overexposure to the whims of nature. And in this endeavour for change, the leading director is—quite naturally—techno-evolution.

But what actually are the driving forces behind all this development? The answer to that question will be discussed in the following chapter.

Chapter 6
Consumption—A Primordial Force

Abstract Today, it is a more or less commonly accepted truth that we who live in the developed countries consume too much and that this destroys the environment. The critics suggests that the only solution is a changed lifestyle—and, above all, reduced consumption. In my opinion, however, a forced reduction of consumption would be a worthless and even counter-productive solution, both for the people of the world and for the environment. Consumption is a necessary primordial force because it is a main force behind techno-evolution.

I am firmly convinced that we are approaching an increasingly sustainable society with rising living standards globally.

I claim that this is the natural unavoidable course of evolution. A movement that in itself should be regarded as a resource.

Arguably, there are also other opinions to explain how the world is developing. One such opinion is that the environmental problems of today are the result of an unsustainable orgy of consumption; the depletion of the earth's resources is so devastating that we would actually need several planets to maintain our continued existence.

Consumption is indeed a powerful phenomenon, a necessary element in both techno-evolution and the growth of civilisation. Figure 6.1 is a graphic representation of the way that consumption drives and expands the technosphere, creating rising living standards along the way—which, in turn, further stimulates higher consumption, technological progress and even better living standards.

It is by no means an exaggeration to claim that civilisation would never have existed without consumption. Nevertheless, many people have developed an instinctive aversion to the concept as such.

Today, it is a more or less commonly accepted truth that we who live in the developed countries consume too much and that this destroys the environment. The critics suggests that the only solution is a changed lifestyle—and, above all, reduced consumption.

EXPANSION OF THE TECHNOSPHERE

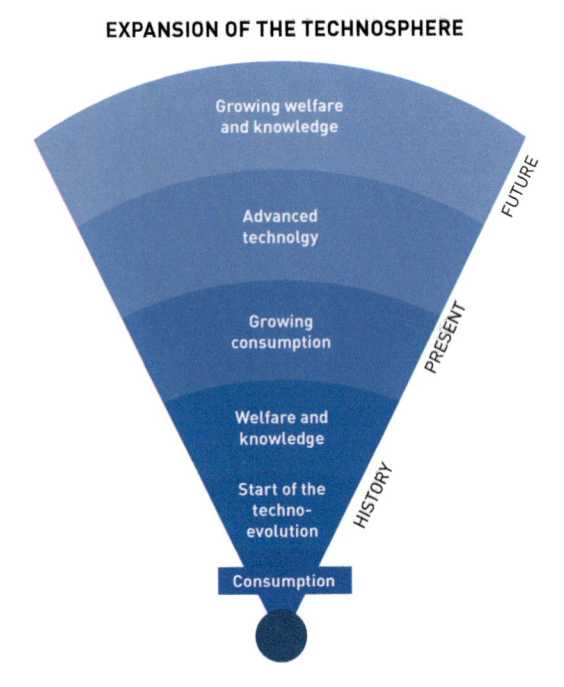

Fig. 6.1 A graphic illustration of how the consumption drives and expands the technosphere, which makes the welfare grow. This will, in turn, stimulate the consumption and lead to even more advanced technology and higher welfare

In my opinion, however, reduced consumption would be a worthless and even counter-productive solution, both for the people of the world and for the environment.

Consumption is a necessary primordial force, which we will now observe in more detail.

Creative Consumption

What is the actual meaning of the word "consumption"? A common definition is: "when you buy something". Another is: "buying, consuming and disposing", or "wearing out and throwing away". "Finishing off" is yet another suggestion.

Surprisingly, most definitions found include an element of ruin; they express the destructive use and disposal of something. It is rather strange that we use such negative words for an activity that mainly involves something entirely different. All consumption has a reason, and that reason is nearly always creative rather than destructive. "Use" or "exchange" would be far better synonyms to use instead.

Consumption, in the context discussed in this book, isn't just a question of consuming resources. It is viewed as the flip side of the coin—the most important one—in the sense that we receive something in exchange. Nevertheless, despite the widely accepted negative connotations, the term "consumption" is used in this book since I believe that it is far better and more important to publicise the complete meaning of the concept rather than to try and introduce a new word.

It is important to keep in mind that today only a fraction of all manufactured goods are disposed of for good; around 10% in the developed countries. This means that most of the consumed products do not end up on the scrapheap; they are re-used, recycled or re-processed to produce energy. We are clearly approaching a recycling society, or a "circular economy", as it is called today. Many European countries have already gone far in this area, and old landfills are being excavated in the search for valuable materials, so-called "urban mining".

Even the words "waste" and "disposal" may one day become archaic terms, at least in the developed countries.

All life—everything from bacteria to people—needs to consume something in order to survive. To fill the stomach, keep the cold out, cure illness and ensure protection, various kinds of consumption are necessary. All activities, from the most basic to various kinds of entertainment, include the consumption of matter and energy. Consumption is about satisfying the need for heat, nourishment, comfort, defence, entertainment, knowledge, security—you name it!

Humans exist thanks to an evolutionary process where consumption has played a decisive part. Archaeological findings have proven that ever since the oldest of times, there were other reasons for consumption than mere survival. Objects like pearl necklaces and various adornments show that humans have always wished to achieve something more than merely satisfy their most basic needs.

Already in ancient times, there was a need to develop alternative solutions for different tasks, like extracting medicine from herbs, finding and processing food, building huts, killing animals and creating fire. People that were good at making tools and other things were asked to make several copies. You could trade a stone axe for something else, maybe some edible roots, a handful of herb medicine or a beautiful, glittering piece of stone.

This early barter and specialisation were important prerequisites for the development of production and consumption. The hominids had the same urge that man still has today—to improve living standards for themselves and their kin.

Consumption Research with a Twist

Psychologists and other behavioural scientists have presented many explanations on why we consume the way we do. A common view—shared by virtually everyone taking part in the environmental and climate debate—is that the consumption of today is unsustainable. This view also has a great influence on the explanations describing consumption behaviour. It is not uncommon that the view of humankind is tainted by negative undertones.

Some of the most frequent explanations for consumption behaviour include the fact that we want to augment our status, express our personality, or demonstrate our personal success. Or that we simply want to stand out from the crowd and receive dopamine kicks. Others suggest that we consume to avoid boredom or existential emptiness, that we struggle for social dominance or that we are driven by envy. These explanations are common both in the critique against an ecologically unsustainable lifestyle and against the present economic system as such.

We have no reason to show disrespect for the behavioural scientists or to doubt the results of their research. All the reasons mentioned above are probably true for some people. My point is not that the behavioural research results are wrong, but that they are incomplete, and that the interpretation of them is misleading when applied to consumption critique.

I suggest that the most common reason for consuming products is that we want to improve our living standards and satisfy our needs, whichever they may be. If a person buys a self-propelled lawn mower, the reason may simply be that he or she doesn't feel strong enough to push a manual mower by hand. So why do we have to invent the reason that the guy next door is already showing off with such a flashy machine?

Recently, I bought an electric toothbrush to get cleaner teeth, but I have never checked what my friends use. I did not receive a dopamine kick or an increased feeling of personal accomplishment.

Now, after having used it for some time, I notice the difference and realise that I should have bought that toothbrush a long time ago. My dentist was right!

It is very unfortunate that such simple basic reasons for consumption are seldom expressed by the behavioural scientists active in the environmental debate.

Another reason for human consumption (that should be obvious) is the need to create. This is an explanation I have never found in any of the books I have read on the subject. Consumption is often about finding ways to create a better life, or the result of a creative impulse born out of plain curiosity, or simply something that provides an inner satisfaction. For many people, shopping for clothes, objects of beauty, furniture etc. is also a kind of creation. We decorate, enhance, refine, and find new shapes and solutions. The playground for the creative mind doesn't have to be a piece of canvas—it can be your own body, home or garden.

The Right Diagnosis for the Right Cure

Viewed in a longer time perspective, and from a functionally deeper point of view, the concept of consumption deserves a much more valued position than it occupies today. While it is true that consumption creates negative effects for the environment, we must differentiate between symptoms, diagnoses and remedies in order to avoid misleading generalisations.

Sometimes we hear things like "the car is the worst threat to the climate". But exactly what does that statement mean? It includes a diagnosis that the car is one of the basic reasons for climate change. Subsequently, "reduced car driving" or "fewer cars" would seem to be the obvious remedy. But the statement that the car is a threat to the climate is the wrong diagnosis for climate symptoms, and therefore the remedy of reducing car driving or the number of cars is also wrong.

A more correct diagnosis for the reason behind climate change would be "the CO_2 emissions from cars". That is, the problem is not the car as such, but its emissions. That is a very important distinction.

As the diagnosis now pinpoints the CO_2 emissions rather than the car itself, the remedy could be something other than reduced car driving. The prescription, based on the correct diagnosis, would then be to reduce the CO_2 emissions of cars. That could be accomplished by a medication that furthers, for example, electric cars powered by solar energy, or hydrogen cars, with no CO_2 at all.

On the other hand, if we settled for the prescription "reduced car driving", fewer cars would be manufactured and bought in the future. But then the development of car technology would lose momentum, and we would have to wait much longer for more environmentally-friendly cars to appear. Some people may believe that that doesn't matter, just as long as car consumption is reduced here and now. But if the development is stalled now, the car technology of today is the one that will be distributed all over the world in the foreseeable future.

In the year 2000, there were 38 million private cars in China; in 2014 there were already 160 million. Still, the part of the Chinese population owning a car is around 11%, to be compared with South Korea (28%) and Sweden and Finland (46%). The future increase in the number of private cars in countries currently approaching developed status is expected to be enormous. With 800 million new car owners arriving until the year 2040, it would be beneficial if electric cars, hybrids, hydrogen cars, and more efficient combustion-powered cars were developed as far as possible. This is something that we in the developed countries of today can support through continued car consumption.

Why is the car so often described as the prime environmental culprit?

This is probably because the car still suffers from being regarded as a symbol of extravagance. And the critics who demand reduced consumption as a solution for climate and environmental problems regard both excessive and luxury consumption as the basic problem.

Many critics associate climate and environmental problems with so-called overconsumption. Among the more well-known, we find the British economist Tim Jackson, the Canadian writer Naomi Klein, the WWF, Global Footprint Network, the organisation Buy Nothing Day, and various societies for nature conservation, together with a large number of scientists and debaters.

One thing that all these critics have in common is that they spend a lot of their time and energy describing the symptoms in great detail. However, when it comes to diagnostics, they all too often go astray.

A frequent suggestion to put the world right is to stop using fossil fuels. This idea is of course basically sound, but once again it has its origin in a misleading diagnosis, namely that energy production using fossil fuels is increasing the greenhouse effect. In my opinion, the correct diagnosis is that energy production *with CO_2 emissions* is increasing the greenhouse effect. The first diagnosis is focused on fossil fuels, with the conclusion that such fuels should not be used at all. In the second diagnosis, the focus is on the CO_2. The difference between the two is critical. Actually, since fossil fuels have the potential to be used *without any greenhouse gas emissions*, why eliminate that possibility entirely? It is already technically possible to separate the CO_2 and store it underground. This technology, so-called CCS (Carbon Capture and Storage), has been used on the Sleipner gas field in Norway since 1996.

The situation becomes even more complex when you realise that the reasons for environmental problems vary from country to country. In addition, the solutions for one country cannot always be directly transferred to another. A fundamental step in all problem-solving is to divide a complex problem into several parts. Overfishing in one part of the world has to be managed in another way than as a global problem. A limitation of car traffic in Scandinavia won't help combat car exhaust fumes and bad air in Chinese cities. Therefore, campaigns to reduce the number of cars in the world in order to get better air quality won't solve the actual problem.

The critique of the consumption society is almost always generalised, and it is often coloured by opinions regarding the redistribution of wealth. There is a strong desire to emphasise how some people living in abundance are responsible for global overconsumption, while others can't even afford the most basic necessities.

An entirely different point of view is that consumption actually empowers growth and the distribution of welfare, even in the world's poor countries. This is, as we know, an old and ongoing controversy. Unfortunately, the climate and environmental questions cannot wait for a consensus to be established concerning the world's economical redistribution politics.

The Luxuries of Yesterday Become the Commodities of Today

Today, very few people would regard a bicycle as a luxury item. On the contrary, the bicycle has become a kind of symbol for an environmentally-sound lifestyle. As has been the case for many other products, the bicycle has been transformed from a ridiculed luxury contraption to an indispensable everyday aid.

The first bicycles consisted of a frame with two wheels and a pair of handlebars for steering, but they lacked pedals and any kind of driving mechanism. The cyclist straddled the frame and simply kicked the vehicle forwards with the feet. The early users were young men and boys from well-to-do families, who became archetypical representatives for their class when they kicked themselves away along the streets of Paris in the 1820s.

Today we should thank these "spoiled brats" for the popularity of the bicycle, as no manufacturer would have continued the production without customers.

For the same reason, the people buying expensive or extravagant electric cars today are a very important consumer group. They started the natural consumption selection that will provide us with cheaper and better electric cars in the future. Because of them, the electric car will probably replace the diesel- and petrol-fuelled cars of today; and the sooner, the better.

The management group of Tesla, the electric car company, realised that in order to attract early consumers with the necessary purchasing power, the electric car had to offer performance and comfort on at least the same level as the most fashionable fossil-fuelled cars. It is still "spoiled brats"—mostly wealthy Americans and Norwegians—who buy the cars and thus drive the technology development forwards.

Rich men of the Western world are commonly regarded as the least climate-wise people of all. Strangely enough, however, these men have probably, through their behaviour, contributed more to a decrease of CO_2 emissions than all the people using bicycles for climate reasons. That is not to say that bicycling isn't good for the climate. It is the critique against consumption which is the paradox in this context.

The Tesla company has been an important pioneer who cleared the way for simpler electric cars, which are more attractive to the majority of customers. The vice president of GM, Bob Lutz, has confessed that he was inspired by Tesla to make GM continue the development of electric cars.

Today, the majority of established car manufacturers have included electric cars in their product ranges.

The mobile telephone has a similar history.

Those who remember the 1980s will also recall the first mobile phones, who were then regarded as symbols of the "yuppie" generation—the newly rich young urban professionals. It was mainly this category of people who could afford—and found it worthwhile—to buy these prohibitively expensive gadgets. (In Sweden, the mobile phone soon become known as the "yuppie-nalle"—a yuppie's teddy bear.) However, these early buyers, who were then widely regarded as bragging luxury consumers, played a decisive part in the future development of the mobile phone.

Today, the results are there for all to see: the production volume skyrocketed and prices went down.

Nowadays, even people in very poor countries have mobile phones. There is probably no other single innovation that has improved the standard of living in poor countries as profoundly as the mobile phone. It may be difficult to realise that this tremendously successful device started out as a luxury item, a toy for the rich, when we now observe the enormous benefits that it has created in the poorest regions of the world.

SMS messaging, for example, can be used for all kinds of communication, from farmers' business calls to healthcare information for pregnant women and first-time mothers. It is also a convenient way to manage your bank affairs, which is a major improvement for people who earlier had to walk or travel by bus for hours to visit a bank office far away. Mobile phones save a lot of time and also further democracy where voting via SMS is now possible.

In some countries, Afghanistan among others, police officers now receive their salaries via the mobile phone. This practice has reduced corruption, because no cash is distributed via corrupt intermediaries. Surveys have shown that among female mobile phone users in the developing countries, over 90% feel safer when carrying a mobile phone, and 85% report that their freedom has increased. More than 40% of women are already using this technology to increase their income and employment opportunities.

History has shown that what is regarded as idle technological snobbery today, can be turned into everyone's benefit tomorrow. Buying a new tablet computer just because you want a slightly updated model isn't as unnecessary as it may seem, because it drives the technological development forwards. The opposite standpoint —no one cares about the updates and no one will buy them—means that the consumption, and thus development, will grind to a halt.

The Involuntary Luxury Consumer

Ecological food is almost always presented as environmentally beneficial, and the advantages of such produce are not questioned by consumption critics. Nevertheless, ecological food is mostly consumed in the rich countries. Poor people all over the world have to struggle to get any food on the table at all.

In the rich countries, it is mainly the wealthier part of the population who buy ecological products, as these are often at least 30% more expensive than conventional foodstuff.

Eco food, earlier known as biodynamic food, has a long history, but it is in the 21st century that the demand for such produce has significantly increased. In 2005 only a few percent of all the foodstuff sold in the Nordic countries was eco food, and in 2015, it was still under 10%. Price is an important factor in purchasing decisions, but the consumers' awareness and confidence in eco food products are also strong incentives.

There are three main arguments for eco food: the food is healthier because less insecticides and artificial chemicals are used, the animals involved lead a better life, and the local ecosystem is less affected. However, the definition of "ecological" is somewhat fuzzy.

For my part, I buy ecological food mainly because I *believe* that it is healthier for me. I emphasise the word "believe", because the scientific evidence isn't unequivocal. On the other hand, this consumption choice of mine has nothing to do with the climate question.

When regarded as a product, ecological food behaves exactly the same way as all other products. The consumption has been limited at the start because of the high price, but as the demand increases, the price relative to traditional products will decrease in a downward spiral. Greater volumes and turnaround allow for rationalisation in the various stages of the production and distribution processes.

At the same time, the competition grows stronger as more producers want a share of the profits, which accelerates the development further. This is product development in small evolutionary steps, which drives prices down while maintaining or even improving quality.

At the moment, however, ecological farming produces less yield per hectare than traditional agriculture, which is a drawback with regard to the increasing requirements of the world's growing population. The yield per hectare ratio will, however, be improved in pace with the development of ecological cultivation methods.

I have based this assumption on the experiences gained from all other production and product development arenas in society. Where freedom rules, techno-evolution will unavoidably occur. If the consumers show that they prefer eco food, traditional agriculture will gradually also be developed in this direction, in order to better fulfil the consumers' choice. Traditional methods will be redirected towards the ecological alternatives.

It is possible that future agricultural life-cycle analyses (LCA) will show that synthetic fertilisers are actually better for the environment than natural ones and will consequently be accepted into the context of ecological agriculture. If so, it wouldn't be surprising. Synthetic fertilisers are easier to customise for specific crops and soils than natural fertilisers.

Increased precision makes agriculture more sustainable, which several scientists have stated. This is accomplished by introducing new technology, for example nitrogen sensors and control systems to calculate the right dose of fertiliser at the right time and place. These methods are already being introduced at many larger farms. In this way, the yield is increased while less surplus fertiliser escapes into the environment. Thus, the production of ecological foodstuff doesn't constitute a return to a simpler, more traditional kind of agriculture. Simplicity and small-scale production do not necessarily mean that the farming is ecologically sustainable.

The example of ecological foodstuff shows once again how important the early (and usually affluent) consumers are for certain products to be established and developed. It is the early adopters that start this important sub-process within techno-evolution—the natural consumption selection.

Active Consumption Supports Sustainable Development

When you stand in a shop, with a package of locally produced meat in one hand and a package of imported meat in the other, you have the possibility to make an active choice. This "active choice" is actually nothing new, it is simply consumption with consideration. An active choice is what you do, more or less, every time you buy a product or service, based on a lot of different selection criteria.

The concept of "active choice" is often associated with ecological or ethical aspects. But criteria that aren't directly based on such considerations can still result in fairly good decisions. Therefore, I prefer to use the term *active consumption*.

From my point of view, active consumption on the personal level doesn't necessarily mean minimising your consumption, rather it means to consume according to your own requirements, and with consideration. And then I mean a consideration that is not limited to ecological selection criteria.

An active consumer drives development in various ways. On one hand, the active consumer, consciously or not, becomes an *early adopter* or *pioneer*; on the other hand, the consumption choices send important signals back to the producer. As I see it, the solution for most challenges associated with the climate and the environment is *not* to reduce consumption by limiting comfort, experiences, education, work efficiency, security, quality of life, and so on. Notwithstanding, consumption will always be limited by your personal economic means in the end.

In other words: Don't stop driving—instead, change to a car that is better for the environment. Travel as much as you want and can afford. Don't shiver with cold in your home but raise the temperature to a level that is comfortable for you—but use efficient heating technology with low emissions if that is within your control.

The statement that active consumption supports sustainable development always raises questions, and I realise that many people may find such a concept horrendous to consider. Shouldn't active consumption by necessity lead to a depletion of natural resources, mountains of waste and increased emissions?

No, the fact is that we can reduce future emissions by consuming today. There are lots of products where each new generation has a lower total energy consumption than the previous, relative to the work accomplished.

The examples are countless: refrigerators, cars, lawn mowers, indoor lighting, washing machines, etc. What these products have in common is that they require more than manual energy to work. Another category includes non-consumer products, like ships, planes, trains, power plants, etc. In other words, all machines, equipment and plants that demand energy for their function or support. Consuming products from these two categories is more sustainable than not doing so.

Interior lighting can serve as a good example in this respect.

Suppose that you replaced all the old light bulbs in your flat with new low-energy bulbs a couple of years ago. That was good for the climate, because the low-energy bulbs demand less electricity. The next year, you discover that there is yet another generation of lighting devices, called LEDs (light emitting diodes), that are even more efficient; they require 20% less energy than the low-energy bulbs you purchased a year ago. The question is: should you replace the bulbs again or just let them be?

Let's imagine that the lighting in your household consumes 1000 kWh of electrical energy each year. With the LEDs, it will be only 800 kWh per year. Even if we consider the energy required for the manufacture and distribution of the LEDs, say a total of 80 kWh, a simple calculation shows that they will still begin to reduce the energy consumption of the household after just 4.8 months.

Therefore, the answer is that it is more sustainable from a climatic point of view to throw away the almost new low-energy bulbs and buy LEDs instead. (Of course, it is an even better solution to give the old bulbs to someone who still needs them).

At this stage, you may also ask yourself if it would be even better to keep the low-energy bulbs and spend the money on something else that will reduce the emissions of greenhouse gases even more. It may be so in reality, but it will still lead to the same conclusions regarding positive climate effects, which is really the point here.

Another kind of consumption that reduces CO_2 emissions is to replace an electrical heating system with one that uses geothermal heating. However, while this reduces the emissions during operation, it also increases the consumption of natural resources for the installation of the system. A geothermal heating system is constructed of mechanical and electronical components, plastic tubes etc. which require use of the earth's non-renewable resources. One issue is improved (the greenhouse gases) but at the cost of another (the use of finite resources).

If we also consider the indirect CO_2-effects created by an increased resource consumption, the choice becomes even more complicated. However, I still believe that most environmental scientists would agree that a change from a greenhouse gas generating system to renewable geothermal heating is preferable, as long as we cannot produce enough electricity entirely without greenhouse gas emissions.

To sum it all up, active consumption is especially beneficial if it ensures that CO_2 emissions are reduced in the long term. Therefore, it is sustainable to introduce new product generations and models that result in less greenhouse gas emissions than to use older products and solutions.

Accommodation, Beef & Car—The ABC of Personal Climate Consideration

Is it climate smart to reduce meat consumption? This question is related to the role an individual can play in lessening the burden on the planet's resources.

It is estimated that 5% of global greenhouse gases are emitted from ruminating livestock, where the contribution of Sweden and Finland together is barely measurable in the global perspective. These countries' CO_2 emissions from ruminators corresponds to approximately 0.02%.

Even if these countries could reduce their part by half, the improvement would hardly make any difference on a global scale. The question is rather if the Nordic countries, small as they are, could serve as good examples for other nations and show the way to a sustainable meat production.

Here there are two possible courses of action. One is to reduce overall meat consumption and hope that this lifestyle will win increasing acceptance in other countries. The other is to develop a more environmentally-friendly meat production and to make the associated knowledge and technology known in other parts of the world.

Sweden and Finland are among the world leaders in this technological field. In Finland, silage serves as an example of how crop rotation and ground structure can be affected so that the cultivated ground emits lesser amounts of greenhouse gases. In addition, most cattle are bred for both meat and dairy production, so we also receive dairy-based foodstuff without any significant additional greenhouse emissions. Another way to compensate for emissions is to use the manure for the production of electricity and bio fuel.

According to the industry organisation Svenskt Kött (Swedish Meat), meat produced in Sweden generally has a 30–40% lower emission level than meat imported from South America. The Swedish Environmental Protection Agency states that the emissions of greenhouse gases from Swedish agriculture have been reduced by approximately 14% since the years 1990 and will be reduced by an additional 20% by various targeted actions during the next 10–15 years.

Countries like Finland and Sweden import several times more meat and meat products than they export. It should be the other way around. We should increase our meat production and export to contribute to a global emission reduction.

However, this would of course increase the local emissions in Finland and Sweden, which illustrates how national environmental ambitions can conflict with global objectives. This paradox is by no means limited to the meat industry.

There is a lot to be said for a continued development of meat production in the Nordic countries. Additionally, these countries have a lot of consumers with both the will and the economy to choose slightly more expensive ecological foodstuff.

Consumption-based evolution is usually better than a boycott. It is better to develop than to discontinue. In my opinion, this is also true for meat production. In other words, a boycott against meat products would be worse than to choose the most climate-smart of two variants on the market.

Boycotting meat will ultimately result in discontinued production. This may of course reduce the related problems, but only if the boycott is effective all over the world. However, if it only affects the countries which have the best ability to guide product development in the right direction, the rest of the world will continue to produce meat in a way that harms the climate.

Therefore, I believe that a changed behaviour that is focused on a climate-smart choice of products will send stronger signals to the market than an outright boycott. It would probably improve the global situation better than to stop eating beef altogether in the Nordic countries.

If I dare to make a prediction concerning the future for meat, I believe that the greatest reduction of emissions will be realised through the development of the products and the production processes, all the way from animal husbandry to the frying pan. But that implies continued meat consumption and an active choice of the most climate-smart products.

An increased number of livestock may actually present itself as one of the most important solutions for climate and food shortage issues, as the expansion of the desert areas may simultaneously be halted.

Too good to be true? Not according to the internationally acclaimed ecologist Allan Savory from Zimbabwe, who has shown that an increased number of livestock and planned grazing can transform deserts into green areas. Even if just a part of these plans will be realised, we might have to think along new lines regarding the issue of meat contra climate.

Buy a New Fridge and Save the Planet

Contrary to popular opinion, it is not necessarily beneficial that products have a long lifetime. If a fridge consumes a lot of electricity, for example, it is a good idea to replace it with a new and more energy-efficient one. In this light, consumption has never been more important than now, since what we consume today will result in more climate-friendly products in the future. And this is even more important as the number of people buying fridges and freezers will skyrocket in the coming decades.

It is estimated that the global number of households buying a fridge/freezer in the next 10–15 years will reach a mind-boggling 500 million, the majority being located in Asia. If all these households bought appliances based on older technology instead of modern, energy-efficient devices, at least 50 more nuclear or carbon-based power plants would be needed to operate them. The development towards more energy-saving cooling technology is still ongoing.

This comparison shows the importance of having energy-efficient products available when the real spending rush hits the market. The more developed the fridges are, the better the chances of tackling global climate change and managing the earths' resources. Everyone can contribute to product development through active consumption. Scrapping old equipment and buying new not only reduces our personal energy expenses and CO_2 emissions, it also reduces greenhouse gas emissions on a global perspective.

Of course, this isn't only about fridges—virtually all products will become more environmentally-friendly in pace with the demand for them. The washing machines of the past consumed more energy than the ones manufactured today, and the machines of tomorrow will be even more energy-efficient.

If we want to see more of this issue expressed in figures, the car may once again serve as an example. Consider the difference in CO_2 emissions if 100 million cars in the year 2030 emit 120 g or more of CO_2 per kilometre driven, as most cars do today, or only 60 g, as the future cars will. A reduction by half is quite plausible, considering the development rate during the last years. That would be a difference of 54 million tonnes of CO_2 per year—approximately the same as the yearly emissions from a country like Sweden or Finland.

Naturally, with electric cars powered by renewable electricity, the difference will be even greater.

Change Your Car—Change Technology

There are two basic opinions regarding the electric car. The critics state that these cars are no better than petrol and diesel cars, because the production and disposal of an electric car consumes no less than 50% of all the energy the car uses during its lifetime. From this point of view, the electric car isn't an environmentally beneficial alternative.

But the supporters of electric cars are not discouraged. Instead, they point out the advantages of electric cars on the roads, where the CO_2 emissions are zero.

In my opinion, both opinions are right.

It depends on two things. On one hand, the type of energy used when charging the car's batteries; on the other, the energy consumed in the factory where the car is manufactured. If the electricity is generated by fossil-fuel power plants, the critics are right—the electric car isn't any better from a CO_2 perspective. But if the electricity exclusively originates from renewable energy sources, then the electric car is a clear winner.

In reality, no country in the world uses 100% energy of either type. In some countries, the energy mix is such that the electric car appears to be no better in an analysis of the emissions of greenhouse gases during the car's entire life cycle.

However, with the mix currently in effect in the EU, the electric car wins. The advantages of electric cars are even greater in countries like Sweden, France, and Finland, because the fossil-fuel power consumption in these countries is relatively low, which can be seen in Fig. 6.2.

Translated into numbers, slightly less than 50% of the EU countries' total electric power is produced by fossil fuel. Poland has a high value, with no less than 90%, and also the Netherlands with slightly over 80%. Finland gets 20% of its electric power from fossil fuels, France less than 10% and Sweden even less, counting the power produced in the country.

The target agreed upon in the EU is that the CO_2-free share of power production should be considerably increased during the coming decades.

Unfortunately, both the time factor and the techno-evolution tend get lost on the way in discussions about eco-friendly cars. The transition to electric cars is a technology shift that cannot be fully assessed here and now.

However, as we have a global agreement on the goal to reduce emissions, we cannot wait to introduce electric cars until all the electricity in the world is CO_2-free. If we start buying electric cars at that point, it is far too late.

Technological development takes time.

To accelerate the production of good and affordable electric cars, the sales volumes have to grow, which will take considerable time. Therefore, it is quite acceptable to choose an electric car already today, regardless of the present energy mix in the power network.

The technical development and growth of electric cars will hopefully proceed hand-in-hand with improved power production.

CAR CO₂ EMISSIONS DEPENDING ON ELECTRICAL SOURCE
Connection diagram – emissions from driving and manufacturing

[Coal based | In the EU | In the Nordic counries | CO₂ free electricity]

■ Petrol and diesel car   ■ Electric car

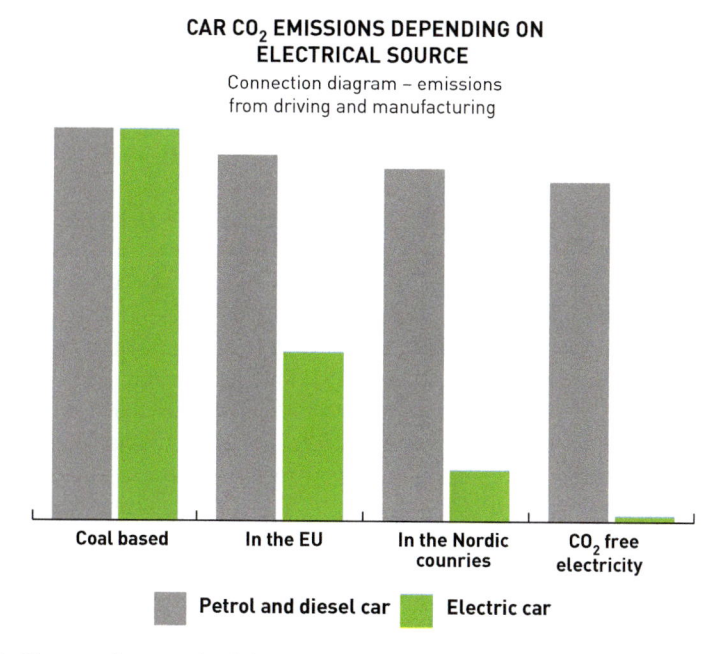

Fig. 6.2 The two bars to the left show that cars running on electricity and fossil fuels, respectively, will be responsible for roughly similar emissions of CO2 if the society's supply of electrical energy is entirely based on coal. The bars to the right show the situation if the production of electrical power is CO2-free. Today, most countries use combinations of different energy sources

The Consumption of Today Creates the Improvements of Tomorrow

What about products whose energy consumption and greenhouse gas emissions are so small that they could be regarded as insignificant? Is it still sustainable to consume new mobile phones, tablets and computers?

My answer is yes. Even though these products require substantial amounts of rare metals and other finite resources, and even though they produce waste and cause CO_2 emissions when manufactured, they are extremely useful. Consuming and replacing such electronic devices, although they may be only a couple of years old, is essential for human development. Despite the ecological drawbacks, mobile phones bring far more benefits in the form of communication and knowledge expansion.

But surely not all gadgets can be justified as meaningful consumption objects? Well, that depends on how you define "gadgets". What some may regard as useless, others consider useful. As time goes by, gadgets tend to evolve into useful or even indispensable commodities.

Some people may regard a GPS as a meaningless electronic gizmo, since you could just as easily roll down the side window and ask someone for directions (as was recently suggested in an environmentalist anthology). But for most people today, the GPS is a very helpful device that minimises the risk of driving the wrong way and helps you to avoid traffic jams, thus saving time, fuel and emissions.

Still, this is only the beginning of all the useful applications to come. For example, GPS-technology, in combination with sensors and smart computer software, can be used for precision agriculture. With data from soil samples, ditch water, humidity and so on, soil classification maps and other basic information can be registered. This knowledge, together with better weather forecasts and continuous inspection of the fields by drones, means that fertilisation can be controlled with exact precision, which will reduce the risk for eutrophication and leaching. At the same time, the use of insecticides can be controlled with higher precision. All this will contribute to making ecological agriculture highly productive in the future.

Buying a GPS supports the development of such technology.

Unlike the products discussed so far, it is also worth focusing attention on a category of products that is seemingly without any vital importance for welfare and progress. We can call them "non-basic products", including various adornments' and jewellery, games, "unnecessary" clothes, and other products and services related to amusement and entertainment. Although these items may represent a lesser value from a techno-evolutionary perspective, their value is still far from negligible.

Clothes are also undergoing development, and are being developed in the direction of higher functionality, new and better materials, environmentally-sounder production methods, less use of chemicals, increased material recycling, and integrated sensors. Most encouragingly, the fashion world has started to act in this direction, not least the Swedish companies. The objectives include the sustainable management of water and chemicals, choice of materials and worker protection in the textile-producing countries. A change similar to the one that was realised in the developed countries some decades ago.

Politicians, the authorities and the fashion industry itself are also introducing quality assurance systems, various kinds of marking and customer information. We consumers will gradually be more able to make wiser selections and further a sustainable development in the textile and clothing business.

Other similar industries are films and video games. The game industry has produced some positive side effects in terms of spreading its technologies into other areas. One such connection is between computer games and body scanning with 3D visualisation in the healthcare industry. This kind of spin-off between different industries is by no means uncommon. It is also another reason why we should be careful before we discard something as unnecessary—we can never know in advance how a new technology may prove to be useful in other areas than it was originally intended for.

Active consumption supports the development of new technology, that in turn may affect everything; and the big advantage is that it furthers the growth of welfare. On one hand, this is because product evolution accelerates and creates better products and services, while on the other hand, consumption creates work opportunities.

Those wielding political power often emphasise the importance of consumption and growth. Hopefully, politicians will keep this point of view and won't let themselves be persuaded to think that consumption itself is harmful for the planet.

It is important to point out that in parallel with active consumption, there is a need for political means of control to ensure a healthy environment. Such means are already emerging in the form of various political initiatives and regulations—emission regulations, demands for energy efficiency, limitations of the use of chemicals, requirements regarding recycling and waste disposal, and so on.

As for the CO_2 emissions, there are already various kinds of consumption that have a direct positive effect. Other good buys—with an indirect effect on CO_2 emissions—include e-readers in place of newspapers, e-books in place of paper books, video games in place of plastic toys, weather safe and stain resistant clothing, etc.

Is Downshifting Possible?

Downshifting is a popular term for the lifestyle change that has often been suggested as a solution for climate and environmental problems. Although there are many suggestions regarding how the changes should be realised, they all include a reduced consumption, preferably in combination with a shortened work week and a simpler way of living.

One of the suggested methods to limit consumption involves economic zero growth or no BNP-related objectives whatsoever. The purpose would be to measure the progress of society in other terms.

Although it is a good idea to use several different parameters to measure society, not only the BNP, disregarding economic growth as an objective is to jeopardise the world's financial and economic systems, with unexpected and potentially unhappy consequences.

There are two basic opinions behind the anti-growth ideology. On one hand, there is the belief that growth (i.e. increased consumption) is the reason for environmental problems; on the other hand, there is a general discontent with the established economic systems, where a redistribution of wealth would be preferable.

However, if we stop the consumption, we will simultaneously stop techno-evolution, and thus man's creative ability to produce useful innovations. With zero growth, the increase and distribution of welfare in the world would grind to a halt, and the result would, in other words, be the opposite of what was intended.

Additionally, we may ask ourselves if we, as a species, haven't already adapted so well to our present lifestyle that we have passed the point of no return. Our requirements have long since been far removed from the basic needs. We are all children of our time.

The Norwegian marine biologist and explorer Tor Heyerdahl (1914–2002) believed that he could live together with the people of a primitive culture under their conditions. He even stamped his watch to pieces to be better suited for living in harmony with the natives in the wonderful South Pacific climate, content with what nature provided. However, in the long run, he couldn't get used to a life without the intellectual challenges, adventures and facilities of modern life. He was "programmed" for a different lifestyle.

In the end, Heyerdahl returned to the developed world and bought a new watch.

Today, there is much talk about downshifting and "getting off the treadmill". But when an opportunity like that presents itself, few people are really willing to take it. The overwhelming majority choses instead to get an education and employment. Most of us want to have a nice place to live, see the world, discover ourselves and achieve a good standard of living. People who have been unemployed can testify how important work is, not only from an economical perspective, but as one of the things that give meaning to our lives. There are also people who work hard, even though their personal finances don't demand it. Maybe we like the treadmill after all?

It seems probable that the best-before date of a simpler lifestyle expired long ago. At least for most of us.

Service Consumption

One variant of the efforts to reduce consumption is to actively consume services instead of products, thereby reducing material consumption. From a climate point of view, the effects of this idea are limited, because most of the CO_2 emissions originate from housing, electricity, heating, traffic, and foodstuff, i.e. from the kind of material consumption that cannot be replaced by services. You simply cannot replace potatoes or shower water with services.

Of course, there are many contradicting examples, where services actually can replace physical products. But in my opinion, an all-out effort to encourage the use of services at the cost of material consumption may actually be counterproductive.

The category of material products affects emissions and the consumption of natural resources the most, and because of that, continued trade and product development in this category is of vital importance. We have to keep in mind that a couple of billion people today are living without TV sets, refrigerators, cars, shower cubicles and computers, and they will be more willing to spend whatever money they may get on products like that than on services. We, the people in the developed countries, are driving the development of these products in the right direction.

Therefore, an increased use of services at the cost of material products will not necessarily support a sustainable development. However, it is quite another thing if a certain function or need can be fulfilled just as well with a service as with a physical product.

Service Consumption

Another political idea to reduce general consumption is to shorten the working week.

Some political parties have suggested that people should work 25% less than today. Then we should take into account that the people in the OECD countries normally work only five days a week, and have five weeks' vacation and a few holidays on top of that. We have paid parental leave and sick days; we spend the first part of our lives receiving an education and can retire before getting really old. This is one of the ways in which we harvest the fruits of the technology-based society we have created.

The idea is that less work and a lower income would permit less consumption, which in turn would reduce emissions and other harmful effects on the environment. It is simple to prove this equation. Unfortunately, it only makes sense in a narrow and momentary perspective.

The people who advocate shortened working hours nevertheless agree that we should install heat pumps, change to electric cars, and shut down nuclear and fossil-fuelled power plants to replace them with renewable energy sources.

We should also build more railroads and develop train technology, help the poor countries to get electrical power, water purification, and more. All these suggestions are good, they represent traditional growth and modernisation. But to realise them, we will need money, lots of money. And that doesn't rhyme with a shortened working week.

Another argument for less work is that people would then have the time to ride a bicycle instead of drive a car, for instance. Some debaters still only buy half of the suggestion though; they would gladly accept shorter working hours but not a lower salary. Such logic, in my opinion, only serves to melt away their climate arguments like the snows of last winter.

As I see it, we should do exactly the opposite—increase the effective working time, if the purpose really is to improve the climate. With increased productivity and income, we can afford to invest in geothermal heating, solar cells and electric cars. Conversely, less money would only reduce the scope for investments in more climate-friendly technology.

Working harder and spending the money on new and more climate-smart products should be a more sensible political strategy for the planet.

I will not delve further into other problems associated with a shortened working work, such as the limited means of financing welfare services, infrastructure, and pensions. Neither will I discuss the question of what people would do with their increased leisure time. However, on a positive note, I could speculate that such a policy is 100 years before its time, since it cannot support a sustainable development at present.

Growth—A Part of the Natural Order

Growth and consumption are two closely intertwined concepts. Moreover, as to whether or not growth should be measured in economic terms is irrelevant; it's all about heated homes, clean water, healthcare, pensions, television and the internet. We owe our welfare to growth.

Because of growth, the majority of the world's present population of 7.3 billion people doesn't have to endure smoke and grime indoors, malaria, lice, and dirty drinking water. In this light, it becomes plain to see that the remaining environmental problems cannot be solved without growth and economic resources. However, the kind of growth needed in the future is one that won't create new and even worse ecological problems.

Many people may find it surprising that peace and freedom are very positively affected by economic growth, which has been asserted by many scientists, among them the Swedish sociologist Bi Puranen, General Secretary of the World Values Survey. This organisation has analysed the development of violence in the world and found that war and violence have been decreasing over a long period of time. Although media reports may seem to indicate the contrary, the number of people killed by war and other acts of violence has actually been decreasing with each passing decade. The scientists associate this positive development with economic growth and increasing democracy.

Perhaps this fact in itself is a reason not to generally condemn consumption and growth, together with its associated trade and globalisation.

Halting the pursuit of economic growth won't solve anything. As growth basically means having a profession and an income, it is therefore an important part of every human's existence. In short, everything that has a part to play in human development is dependent on economic growth.

The suggestions for political regulations to reduce emissions often focus on consumption limitations; air travel being one such example. A lot of people succumb to a bad conscience after spending a holiday in Thailand. They may eventually decide to limit any future travelling in the belief that by doing so, they will be making a meaningful sacrifice for the good of the climate.

Reduced travel will have many consequences, although not all of them are positive. Thailand, together with many other countries, owe a lot of their own growth and welfare to tourism and travel.

The developed countries' consumption of computers, mobiles, and clothing has increased growth and helped hundreds of million people in India to rise from poverty. Without our enthusiasm for consumption, China would be a country without growth, and its population would probably be starving.

The developing countries have every right to modernise and to distribute welfare among an increasing part of their population. But they wouldn't gain anything from reduced global trade and travel.

Even though a limitation of a certain kind of consumption, for example air travel, seems to be climate-smart from an individual perspective, it may not necessarily be smart in a global perspective. Moreover, reduced air travel would slow down the development of environmentally-friendly planes.

Similar repercussions could also occur when focusing political means of control on reduced car driving rather than on the emission problems.

To return to the opening message of this chapter; consumption is a primordial force. It is powerful in several ways, not least for growth. However, consumption will also have some side effects and—exactly as with medicines—we must find ways to minimise those effects by improving the medicines.

Exactly What Is Wasteful Consumption?

Decisions on the types of consumption regarded as extravagant—and the consumers to blame—are often made according to strange criteria. In the desert climate of Dubai, there is a giant indoor ski slope, and it is easy to condemn it as a case of despicable luxury consumption and a gigantic waste of resources. But the same people who criticise the ski slope have no objections to visiting heated public swimming pools in the Nordic countries.

When the outdoor temperature in Dubai is +33 °C and the indoor temperature of the ski slope is −2 °C, the difference in temperature will be 35 °C. In our public pools, the indoor temperature is normally +24 °C, so when the outdoor winter temperature is −11 °C the difference between the indoor and outdoor temperature would be exactly the same, i.e. 35 °C.

The only difference is that we have a lot of heated public pools in our countries, while Dubai has only one ski slope; with a higher utilisation, we may add.

The consumption debate has a lot to do with opinions and values. Fashion is another example of an area often blamed for extravagant and unnecessary consumption. The writer and fashion historian Tonie Lewenhaupt shows in her book *Inte bara mode* ("Not only fashion") how deeply rooted and hereditary the concept of fashion is in the mindset of the creative being that humans have evolved into.

She writes:

> Nothing is more sensitive, more perceptive to the shifting values of society than fashion. No other creative expression is faster to reach the consumers, no other new-born ideal is easier to test, accept and adopt. In the same way as art, music, architecture, dancing, photography, and film is criticised, reviewed and regarded as culture, fashion should also be an obvious part of the same creative world. Even so, fashion has long been regarded as unnecessary, vain, superficial and non-intellectual.

These examples show how different we value consumption. What one individual regards as extravagant, wasteful, unjust, unethical or generally despicable, another may see as an expression of creative and intellectual culture.

At this point, we may ask ourselves if there really is something as totally useless consumption. Although we could hardly answer no to that question, in my opinion, consumption that doesn't fulfil any need at all is quite rare.

However, the consumption of energy that doesn't give any comfort or added value, like "making a fire for the crows", is not useful.

But what about the use of energy that creates emissions? Well, it isn't good but it can still be useful. The bad thing in this case is the emission in itself, not the consumption as such.

Consumption involving human suffering or cruelty against animals is of course objectionable, as is consumption beyond your means. And although these topics are worthy of serious debate, they fall outside of the scope of this book.

However, I decided to challenge myself to think of an example of any product or service that appears to create a lot of emissions without giving humankind much value in return. And one area I thought of was space tourism. A lot of resources and work are spent to allow a few people to experience a short period of time in the so-called Karman line, the boundary between the atmosphere and deep space.

That, at least, must surely be entirely unsustainable?

The answer, however, is that we cannot know for sure. History is full of examples of important things that man has unveiled by travelling beyond the boundaries of the known world. It isn't easy to predict which kinds of consumption will prove to be most useful in the end.

Is Happiness for Sale?

When we fail to find other arguments against consumption, the one about happiness is always at hand. According to this line of reasoning, we cannot make ourselves happier by consuming more. The idea is to show that life satisfaction isn't associated with increased growth and consumption. Therefore, we can gladly reduce our consumption, spare the environment, and improve our quality of life at the same time.

It is certainly worth taking a closer look at this standpoint, as one of the most central theses of this book is that continued consumption is of vital importance for sustainable development.

People who criticise consumption in a happiness perspective usually point out that the general level of happiness hasn't changed in the latest 60 years, even though the BNP and average income have increased significantly in the same period. There are often references to the "Happy Planet Index" (HPI), an attempt to measure the level of happiness in different countries. The calculation involves an evaluation of people's experienced well-being and life expectancy relative to the local ecological footprint. According to the index, this proves that we who live in the richer countries don't become any happier because of increased consumption, as we have already passed the maximum limit of increased happiness long ago. Consequently, we could begin to generally reduce our consumption, with no ill effects.

The absurdity of these conclusions can be demonstrated by the following example: Suppose I bought a tablet PC that provided me with some benefits and fun. If I then took part in a happiness survey, I would probably appear to be slightly happier than before. The tablet PC would have increased my wellbeing, at least a little.

Two years later, I would then take part in another happiness survey. The tablet PC would no longer affect my happiness as much, because now I am used to it. The later survey would therefore show that my happiness level was not affected by material things.

In this way, the happiness researchers are able to conclude that people don't become any happier by material consumption, except for only a short time after the purchase. This would be true for all equipment like colour TV's, refrigerators, tumble driers and everything else that has been made affordable for most people in the Nordic countries since the 1950s.

But let's imagine that all these devices would suddenly be forbidden. Not only would the happiness score fall to the lowest value possible—it would most likely cause a revolution of hitherto unknown fury. And we are not only speaking of the feeling of being robbed and degraded by dictatorial means. No, the main reason would be that no one would want to return to a life without colour TV, modern homes, air travel and the internet. In reality, very few would accept going back in time, not even to the material standards and welfare of the 1990s. This is also evident from migration; very few people are voluntarily moving to countries with a lower material standard of living.

According to the latest HPI index, the top positions were occupied by countries like Costa Rica and Vietnam. Cuba and Bangladesh were awarded the comparatively high positions of 11 and 12, while Sweden and Finland had to be satisfied with the positions 52 and 70, respectively.

Still, there are no waves of immigrants in pursuit of happiness moving to Cuba or Bangladesh. This seems to indicate that the conclusions of the HPI index are rather unrealistic arguments on how we should live our lives to achieve happiness.

This standpoint is supported by other similar calculations, which show very different results. In the *World Happiness Report* of 2015, all the Nordic countries were among the top 10 of the 151 countries accounted for. And in another survey, made by the Office of National Statistics in Great Britain, the Nordic countries topped the list in Europe in terms of general life satisfaction (Bulgaria achieved the bottom position). These surveys confirm that there is a strong positive correlation between a good economy and satisfaction with life.

Happiness estimates and theories about life quality are nonstarters if the purpose is to persuade people to consume less and choose a simpler lifestyle.

Still, consumption critics can claim that the products we either know nothing about or are still to be invented have no value for our present state of happiness. So it is, of course. However, the conclusion with such reasoning is that we would have all been just as happy as *Homo erectus* with stone axes as the only tools. This, in turn, raises a question as to whether there has possibly ever been a time in history where a sustainable level of consumption and welfare was achieved, and when the development should have been halted.

To sum up, it is important to think twice before condemning consumption. In my opinion, we will never be able to solve a debate where different human activities and consumption preferences are pitted against each other. In a larger perspective, it is actually the wrong issue to focus on.

Instead, we should concentrate our efforts elsewhere, on active consumption and technological change forced by well-considered political means of control.

We cannot save the planet by consuming less and going back to a simpler lifestyle. Instead, we have to revaluate consumption and accept it as a positive force.

Chapter 7
Real and Imagined Threats

Abstract Since the focus of attention is so profoundly fixated on the problems in the climate and environmental debate, the progress already made—and the opportunities at hand—are often overshadowed. There is a lot of progress going on in the world today, but not without negative side effects. When improving the world and dealing with the side effects, an optimistic attitude provides us with a much better chance of success than a pessimistic view. There are clear reasons why pessimism does not support a sustainable development.

Threatening reports about our ability to create disasters and even exterminate ourselves are not a new idea. A standard example is the British national economist Thomas Malthus in the early 19th century, who predicted that population growth would come to a halt because of starvation.

Malthus calculated that the available food in the world couldn't feed more than one billion people. He extrapolated the development from a still picture of his own time and couldn't fathom that food production would increase tremendously thanks to new knowledge and technology.

Our present food production is sufficient for seven times as many. Malthus didn't pay attention to the fact that we live in a continuously changing civilisation, and the same kind of miscalculations are still made today.

There are people who have even achieved the status of media superstars by presenting various dystopias and catastrophe scenarios. As early as 1968, Professor Paul Erlichs at Stanford University published the bestseller *The Population Bomb*, where he predicted that an imminent population explosion would result in hundreds of millions of deaths by starvation in the 1970s and 80s. Basically, he made the same mistake as Malthus, i.e. he treated knowledge and technology as if they were static phenomena.

The most widely read environment report in the world, *State of the World*, was a loud whistle-blower when it was first published in the early 1980s. The Swedish version, *Tillståndet i världen*, was published yearly from 1984 and some years into the 2000s by the Worldwatch Institute Norden; I still have some of the early issues left. This report contains many valuable observations and suggestions, but also several basic analytical mistakes. In other words, it acts as an eye-opener, but it suffers from being tainted by political ideology.

Its main weakness is that it doesn't take the intrinsic driving forces of progress into account.

State of the World was translated into most major languages and is, as already mentioned, the world's most widely read environmental report. It has affected us all, directly or indirectly, through school and media. Even if the Swedish version I refer to was written some years ago, it is still worthy of discussion, firstly because it maintains an appearance of scientific validity, and secondly because it has served as a trendsetter for the general ideology which has been adopted by many later books and reports on the subject at hand. It still lives on as an engraved pattern in our conception of the world.

In the report we can, for instance, read the following:

> A world where human desires and needs are fulfilled without the destruction of natural systems demands an entirely new economic order, founded on the insight that a high consumption level, population growth, and poverty are the powers behind the devastation of the environment.
>
> The rich have to reduce their consumption of resources so that the poor can increase their standard of living.
>
> The global economy simply works against the attempts to reduce poverty and protect the environment.
>
> We stubbornly insist to regard economic growth as synonymous with development, even though it makes the poor even poorer.
>
> Even if we up to this point have mainly described the environment revolution in economic terms, it is, in its most fundamental meaning, a social revolution: to change our values.

Massive threat scenarios are still presented, for instance in the British scientist Tim Jackson's book *Prosperity Without Growth* from 2009, which is one of the most widely read and frequently quoted works in this area. Tim Jackson, who is an economist and professor in sustainable development, explains how we humans are indulging in a ruthless pursuit of new-fangled gadgets in a consumption society running at full speed towards its doom.

He also claims that material things in themselves cannot help us to flourish; on the contrary, they may even restrain our welfare. In other words, we cannot build our hopes that the economy, technology or science can help us to escape from the trap of *Anthropocene*, which has brought us to the brink of an ecological disaster.

There are hundreds on books on this theme, and they all agree that the general state of the world is pure misery; everything is getting worse, the resources are being depleted, and that man will soon have destroyed the entire planet. The apparent reason for this, of course, is due to the consumption culture and the present financial system—which exposes man as a greedy, ruthless and ultimately weak creature.

This attitude may serve a purpose as an eye-opener. But it is not very credible, and it may even be counterproductive. Of course, we can see a lot of problems ahead of us; but to solve them, we need the correct diagnostics instead of dubious doomsday prophesies.

Focus: The Problem

Since the focus of attention is so profoundly fixated on the problems in the climate and environmental debate, the progress already made—and the opportunities at hand—are often overshadowed. The example below will help to illustrate this point:

In the year 2014, the Nobel Prize in physics was awarded to three scientists who had invented blue light emitting diodes—a technology that has made high-bright and energy-efficient LED lighting possible. As lighting accounts for 20% of the world's total electrical consumption, this invention has the potential to radically reduce energy consumption and greenhouse gas emissions.

In an interview made by the major Swedish daily newspaper Dagens Nyheter, one of the prize winners, Hiroshi Amano, says the following about energy-efficient, inexpensive and high-bright LED lights: "They are now being used all over the world. Even children in the developing countries can use this lighting to read books and study in the evenings. This makes me very very happy".

Shortly after this announcement, the news headlines declared that LED lighting was a threat to the environment. This statement was based on a report showing that LED lighting could be hazardous to flies and moths, which in turn might disturb the eco system.

This is a typical example of how progress pessimists and, not least the media, think and act. In this case, they focused on a potential problem associated with LED lighting, and ignored the tremendous possibilities that the new technology offered to dramatically reduce greenhouse gases and thus spare the eco system (not to mention all the other advantages).

Books and reports of the kind mentioned above tell us repeatedly about disasters, threats, problems, collapses and famines. On the other hand, they are notoriously silent about the great improvements actually made—the reduction of extreme poverty (not only as a percentage but also in absolute numbers), longer lifespans, dramatic global progress in education and healthcare, etc.

The lack of positive media coverage on the environment means that many people believe that too little is being done, which is quite understandable considering the one-sided nature of the information they are presented with. Alarmist reporting almost always reminds me of pirates: they are unreliable and half their vision is blocked by their eye patches.

It is vital that the media not only one-sidedly focus on the misery without presenting the progress made and suggesting constructive courses of action. The quality of our decisions in all respects depends on our knowledge, insight and attitude.

Real and Imagined Threats

Many people are convinced that the climate and environmental problems are growing. It is certainly true that our planet has its limitations, but many of the predictions from alarmist literature have been proven false.

In the 1980s, the forest dieback was a frequently discussed subject. To quote the well-known German news magazine *Der Spiegel*, an "ecological Hiroshima" was imminent. Most experts at the time claimed that a wide-spread forest death seemed unavoidable. Additionally, the general mood of impending doom was augmented by the threat of a nuclear disaster during the cold war. I remember the pessimistic discussions among friends and how frequently the gloomy reports appeared in Swedish and Finnish television. The future of humankind appeared to be depressingly bleak.

But the forest dieback never happened.

On the contrary, the forest area has been constantly expanding in Europe, even during the entire period when the forest was believed to be dying. Today, only two thirds of the yearly accretion in Europe are cut down, according to the Natural Resource Institute in Finland.

There are different opinions as to why the large-scale forest dieback didn't occur. One theory is that the researchers' evidence and conclusions had been incomplete and too hasty; the forest was actually never in danger. Others suggest that the emission limitations implemented prevented the disaster.

My point is that the environmental catastrophe *did not happen*.

Some other environmental problems, exaggerated or not, that have concerned us during the last decades have also disappeared from the immediate agenda: over-population, DDT, the ozone hole, heavy metals, lead poisoning, soot particles, the waste mountain, and the acidification of our lakes. Unfortunately, some environmental problems, like soot particles and waste, still remain in some areas, especially in poorer countries, where there are other, even worse problems that have yet to be resolved.

The conclusion is, however, that we and our society in most cases have handled threatening situations quite well. When alarming symptoms are noted, scientists and other experts are summoned, and we act according to their diagnoses. It is no big deal that the diagnoses are sometimes wrong, as long as the side effects are not too severe. The main thing is that we do our best to avoid disasters, and on the whole, humankind has succeeded rather well this far.

As individuals, we react very differently to various kinds of threats. The closer and more tangible the threat is, the more violent are the reactions—while distant and invisible symptoms, like the depletion of the ozone layer, concern us less. In the latter cases, we have to trust the scientists' and later the politicians' reactions.

Does this mean that disasters are avoided thanks to war headlines, threats, and anxiety? I don't think that this is the most important explanation; rather, it is factual and science-based information that produces effective results. But if exaggerated threat scenarios and reports of misery are needed to inspire the necessary political opinion, acquire research funding and create behavioural changes, we will have to live with that.

The most important thing to remember in this context is that the actions shouldn't cause more harm than the original problem itself. The risk with exaggerated threat and misery reporting is that it may inspire an over-reaction based on misleading diagnoses, or the opposite—a paralysing feeling of helplessness. It is necessary to take threats against the climate and the environment seriously, but not to a degree where our ability to reason and act is blocked by fear or anxiety.

Many environmental debaters claim that the fall of the Inca and Roman empires were caused by the same causes that are now threatening our present civilisation—a short-sighted over-exploitation and rape of nature. Easter Island is another popular example. However, in my opinion it is both worthless and irresponsible to judge the world situation of today by copying the outcome of earlier cultural endeavours in history. The inhabitants of the Inca empire and Easter Island didn't have anything even remotely comparable with the organisations, technology, medicine or general knowledge of today.

It would be like comparing a case of appendicitis in the past to a case today. In pre-modern times, it was a fatal condition. In this day and age, it is cured by a simple routine operation.

Today, humankind is conscious of the climate changes and other ecological challenges. And we also have the knowledge and resources needed to act.

Facts, Propaganda and Hidden Messages

During all the years I have followed the development of technology and society, I have repeatedly observed how a mishmash of serious research, political propaganda, and the hidden agendas of individuals have been distributed more or less randomly by the media. There are of course many different kinds of alarmism—everything from well-founded research reports to exaggerated prophesies of doom. It is far from simple to separate the wheat from the chaff.

The actions taken against ozone depletion, lead emissions and the toxic chemical, dioxin, are all examples of how research has shown the way to successful results. Today, greenhouse gas emissions top the list of issues deserving our gravest attention, as it is a global phenomenon—just as the depletion of the ozone layer once was.

There are also a considerable number of local environmental problems, such as drought, air pollution, forest depletion and overfishing. All of these are real threats that have to be acted upon, even though they are not global.

However, I am always disturbed when a single global environmental issue is bundled with an assortment of several local issues, rather like a simplified trademark advertisement for the negative consequences of civilisation. This makes the information abstract and inaccurate, ignoring the fact that different locales require different solutions.

Fear and alarmism are natural reactions that once protected us when we were living at the mercy of nature—they are evolutionary relics from our life in the savanna. Today, the same properties can be significant drawbacks.

The transition from a primitive, animal-like state to the society we have today must, on the whole, be counted as a great success. But many people regard the same world as over-exploited, depleted, unjust, war-ridden and balancing on the brink of destruction. How can people living in the same epoch have so entirely different views of the world?

In the sustainability debate, there is one faction dealing with the natural resources and ecosystems, and another focusing on the redistribution of wealth. There is even a third faction discussing a minimalistic lifestyle; for example, downshifting, with less work and less material welfare.

When all these ingredients are mixed without discretion, the result is an anxiety soup that many have choked on. In a situation like that, we cannot expect any constructive initiatives to materialise. Instead, it would be far better to explore, research and discuss each dimension separately.

What Is the Real State of the Planet?

It is easy to generalise and say that we over-exploit the planet's resources and pollute the world with our waste. But how many care to examine these statements in detail and ask exactly which resources are over-exploited?

- *Are fish becoming extinct?* It is true that overfishing occurs in many places, which is, of course, unsustainable. However, this is not an unavoidable threat to the world's total food resources. Fortunately, there are several examples of fish stocks that have either recovered or started to replenish once the fishing effort has been eased.
- *Is the air being poisoned?* Many are convinced that the air we breathe is becoming dirtier all the time. But that isn't true, at least not in the Western world. From the year 1990, emissions of sulphur dioxide have been reduced by 80%, nitrogen oxides by 44%, volatile organic substances by 55%, and carbon monoxide by 62%. Despite these dramatic improvements, 64% of Europeans believe that pollution is increasing.
- *Are the forests dying?* It is a general belief that the forests in the developed countries are dwindling. But that isn't true; on the contrary, the wooded areas are expanding. However, the forests are decreasing in the poor countries, where forestry and farming are still major sources of income, as they once were in the industrialised countries.
- *Are we drowning in waste?* There are many who believe that we are surrounded by constantly growing mountains of waste. In the developed countries, the truth is that increasing amounts of waste are being recycled and the landfills are decreasing.

- *Will there be enough phosphorus?* Phosphorus is an important nutrient in farming, extracted from phosphate ore. Many scientists fear that the finite natural resource of phosphate ore will become depleted in the future, which may jeopardise the world's food supply. But there are already working solutions for this problem, such as by reclaiming phosphorus through digestion residues and sewage sludge. There are also technological solutions for the chemical extraction of phosphorus from polluted water—the remediation of lakes and rainwater by removing phosphorus is already a common procedure. Here we achieve a win-win situation—phosphorus is collected while preventing the eutrophication of lakes.
- *Will there be enough energy to go around?* A common statement is that the earth's population is too large, and that we consume too much energy with respect to the climate. This is one of those issues where we have to think in terms of symptoms, diagnoses, and medication. The symptoms are there for all to see: climate change.

On the other hand, the diagnosis that we consume *too much* energy is wrong. The correct diagnosis is that we are not using the right technology; i.e. energy efficient power production without harmful emissions. Consequently, the correct statement would be that we consume energy that is *produced by technologies that are harmful to the climate*. The difference in wording is important. As the first diagnosis is "too high energy consumption", the remedy will be to use a different medication than a diagnosis based on "the wrong technology".

Alarmist reporting can inspire bad decisions if the statements aren't systematically reviewed and evaluated. It can also be misguiding to express environmental threats in general terms. Actions must be based on precise specific symptoms with corresponding diagnoses. If the doctor discovers that the patient is lame and suffers from a high fever, it doesn't help to predict imminent death. Maybe the lameness and the fever have different causes altogether! A successful cure would probably include two different diagnoses with separate medications.

Several recent surveys of the general conception of the world have been made—one is Project Ignorance by Gapminder and Novus in Sweden. One of the questions asked was whether CO_2 emissions per capita and year had increased or decreased in the world during the last 40 years. The surveyed group was large and representative in order to give a fairly accurate picture of the common opinion.

No less than 90% believed that CO_2 emissions had increased. The truth is that they haven't increased at all.

It is important that decision makers on all levels learn how to see the wood from the trees. Decisions based on false preconditions can halt technological development, and thus also the development of the economy, welfare, and a healthier environment.

The flow of innovations in the climate and environmental areas is accelerating rapidly. This can be seen in the number of improvements that have occurred in recent years, which can be counted in the thousands. Such improvements have to be weighted on the same scale as the problems in this area. That is not to say the problems should be ignored—they need to be acted upon. But they should not be allowed to occupy our brains to the extent that our power to act is paralysed.

Is the Notion of Sustainable Technology-Driven Growth Over-Optimistic?

The development of a technological society has always been questioned. In the 19th century, critics claimed that the technological revolution would create poverty. In the 1970s, it was generally believed that the forest dieback would cause a disaster. In the 1980s, the acidification of lakes and throwaway mentality of society were regarded as manifestations of the devastating properties of growth and industrialisation. Today, many fear the environmental effects of air travel and the production of electronic devices.

There are people who seriously wish to halt economic growth and wind back the clock to the society of the 1960s. They recall this time period as small-scaled and down-to-earth, stress-free and idyllic. But they tend to forget that the refrigerators of that time required 90% more electricity than today, and that our teeth were repaired with mercury fillings instead of plastic. There were no X-ray CT scanners and no medicines against ulcers.

In addition, there were many more people living without electricity. There was also more widespread malnutrition, a higher infant mortality, and, in fact, more wars. Cars were fuelled by leaded petrol, and sulphur emissions were 90% higher than today. The acidification of lakes, as well as polluted streams and fields, were serious concerns.

Since then, technological innovations have reduced sulphur emissions and removed the lead from car fuel.

At any given point in history, there have been critics claiming that this was the time when we had reached the optimal point in the development of the modern society. But we hadn't, not then and not now. And the more our countries are modernised, the greater our possibilities to care for animals and nature become.

In the mid-1800s, the killing of large animals like sperm whales didn't concern people to any significant degree, despite the cruel hunting methods using harpoons. The benefits of the whale fat, mainly used for lamp oil to facilitate reading in the evenings, overshadowed any empathic impulses.

In the 1850s more than 70,000 people were employed by the American whaling industry. There were 900 ships in the world hunting whales, and during one of the most active years, 8000 whales were butchered, which provided more than 300,000 barrels of oil.

The oil extracted from the head of the sperm whale, the so-called spermaceti oil, was especially sought-after. It was of very high quality and sold for 1.50 US dollars per litre in today's monetary value. As a consequence, the number of sperm whales in the world rapidly dwindled.

However, when oil drilling started in Pennsylvania in the year 1859, the price of whale oil began to fall. The fast transition to petroleum products for lighting and other applications is considered to have saved the last of the sperm whales.

Thus, new technology can both contribute to the protection of threatened animal species and provide the wealth to make it affordable for us to even save predators. Imagine what would happen if we were able to bring back someone from the 19th century and tell them that today we move wolves though the air by helicopter in order to save the species and expand its habitat; our ancestor would probably rather go back to sleep than listen to such apparent stupidity.

Pessimism Does not Support a Sustainable Development

There is a lot of progress going on in the world today, but not without negative side effects. When improving the world and dealing with the side effects, an optimistic attitude provides us with a much better chance of success than a pessimistic view.

The optimist carries a positive inner beacon to follow, while the pessimist is always looking for potential traps and drawbacks. As visions and conceptions of ideas often become self-fulfilling, it isn't difficult to realise what's most constructive.

All decisions—big or small, conscious or not—are affected and guided by our inner beacon. When solving a problem, such as developing a new product for example, it is necessary to have a conception of a working solution in mind.

As a product developer, it is of course necessary to review every minute step in the process and question the choices made. You have to ask yourself if there may be a better material or a smarter design. Strange as it seems, this continuous struggle in the mind of the developer may appear to be a kind of pessimism, as it is all about looking for weaknesses in the imagined solution. It is not dissimilar from the process a doctor follows when selecting a diagnosis and a remedy. You start with certain hypotheses, examine, exclude, test, question and verify until you are satisfied that you have made the correct diagnosis.

Then the choice of medication becomes much simpler.

It would be fatal if the doctor was pessimistic from the start and worked in the belief that it would be impossible to find a reason for the illness, or a working remedy. This could then be the conclusion that such a doctor would unconsciously try to verify.

Would you like to have a doctor like that? The same is true for climate and environmental problems—we need optimists armed with critical thinking to solve them.

There are also so-called climate change deniers, who believe that man hasn't really affected the planet and its ecosystems to any significant degree. Some of them claim that the influence of the sun and other natural phenomena are so enormous that human activities have no bearing on global warming. Perhaps these deniers are so deeply pessimistic that they cannot imagine any possible solutions.

For ages, man has harboured a certain distrust of his own species. Throughout history, various religions have emphasised human shortcomings and presented assorted consequential threats. During the last 30 years, such prophesies have increasingly often been introduced by environmental activists and some political groups, whose messages have been significantly supported by the media. The underlying conception of humanity isn't flattering. The human race is considered to be fundamentally ruthless, greedy, short-sighted and evil. Threats against the climate and much other misery on earth are caused by human failure.

However, if we take the time to study the progress that has been made by the human race throughout the ages, we actually get the opposite picture. Can it really be evil, greedy, and short-sighted beings who put their own lives at stake to treat people infected by Ebola or HIV in poor countries? Who are the ones that are continuously reducing the number of starving people on earth? Who are the ones that invent vaccines for the children of the world? Who are the ones that have developed a civilisation where an increasing number of people get educated, and who struggle to reduce the casualties of war?

Why blame an entire species for atrocities that are actually committed by a mere fraction? Establishing a firm belief in humankind should be the first step on the road to sustainable development.

Chapter 8
The Welfare Debt and the Rebound Effect

Abstract Can we conclude that the industrial countries really owe the poor countries an environmental debt? The answer is not self-evident, because the developing countries' ability to increase their living standards depends on the knowledge and innovations that have been created in the industrial countries. Our ancestors bathed two times a year, suffered from lice and coughed perpetually in their huts due to the smoke from the wooden fires which was harmful for their lungs. Our generation fares incredibly much better, despite some environmental issues. The debt to future generations will be worth far less than the heritage they will receive, i.e. all the accumulated knowledge, medical solutions and useful things that have been created up to this point in history. In this context it would be meaningless to speak of a rebound effect, because it is actually nothing other than an improved standard of living over time.

In the mid-1990s, my work brought me to Kathmandu in Nepal. There I had the honour to give a speech regarding modern biological waste treatment to the Minister of Health and the professionals responsible for sanitation. The audience was enthusiastic about our biogas technology and wanted to have it installed in Kathmandu.

Regrettably, despite all the enthusiasm and attempts to engage various aid programmes, we were unable to raise the funds needed to start the project. The waste strategy in the country, one of the poorest in the world, is still based on landfills. Fortunately, the waste is no longer dumped in the river that also provides the drinking water.

From experiences like this I have learnt that poor countries will not prioritise environmental problems as long as they have even more pressing matters to attend to. When there isn't any money for things like basic healthcare, environmental projects will appear far down on the priority list. Therefore, I am convinced that growth and technological development is the only way forward.

On the other hand, we should keep in mind that the situation isn't hopeless.

In the 1960s, no one would have dared to predict that the world population would have doubled in 2010, while at the same time less people would be starving than ever before.

Another example: when the Millennium Development Goals for the year 2015 were defined in 1990, many people regarded them as a utopian dream. For example, how would it be possible to cut the number of starving people in half in just 25 years? But the follow-up documentation *The Millennium Development Goals Report* from FN/UNDP clearly shows that the development has succeeded beyond expectations.

For example, the number of people living in extreme poverty has been reduced by 700 million—at an even faster rate than what was originally planned. Between the years 1990 and 2012, no less than 2.3 billion people were provided with clean drinking water, thanks to purification plants. Also in this case, the original objective was surpassed.

However, much remains to be done, and the differences between regions and nations are vast. Therefore, 17 new global development goals were decided upon in 2015, to be fulfilled in 2030. The challenges against a sustainable development are many, and the outcome will be strongly affected by economies, political situations and conflicts.

As for education, the development looks promising, and there are strong indications of a leap in general knowledge outside of the traditional school systems. In this field, computers and smartphones will once again play an important role.

According to UNESCO, a literacy revolution has started in the developing countries. The price for information and learning aids has fallen, while educational material is being developed and made more accessible by technological means.

Does an Accelerating Development Increase Inequality?

Many people are convinced that they enjoy their welfare at the expense of others. After all, why aren't the benefits of our technology and welfare made available in the poor countries? These are unspecified, generalising and common questions.

A more relevant question would be to ask why the improvements aren't distributed at a faster rate. One answer is simply that it takes time to eliminate the problems, in some cases decades or even up to a century. Industrialisation and modernisation were started at different times in different parts of the world. Some countries went through this revolution as early as 200 years ago, while others are still primitive agrarian societies.

Other reasons include a lack of reform—both political and religious—as well as a lack of democracy. Of course, military conflicts are also difficult obstacles on the road to a modern society.

Unfortunately, inequality will probably continue to increase for a while, as the modernisation process accelerates. This may seem strange, but I will try to explain why.

Whoever watched Usain Bolt win the 100 m sprint in the 2012 Summer Olympics must have noted his exceptionally powerful acceleration. He kept on accelerating for approximately 65 m from the start, at which point he had achieved his maximum speed. If an identical copy of Bolt had started two seconds after the starting pistol had fired, there would have been a clearly visible expanding gap between Bolt and his clone. This gap would have widened until the moment Bolt finished accelerating and ran the rest of the distance at an even speed. Thus, the distance between them would have *increased* even though both were running, were equally talented, and executed the same performance. It all depended on when they started. Once again, a snapshot wouldn't reveal the whole truth.

Industrialisation and welfare accelerate in a similar way. The participants who start later will have to see the gap between them and the forerunners increase in the beginning. A sprint can never be fair if the runners don't start at the same time. The difference in industrialisation and the associated prosperity between rich and poor countries can partly be explained by the fact that the richer countries started their industrialisation process earlier.

Despite the inequality that is created by this accelerating phenomenon, the state of the poor countries would have been even worse than today if there hadn't been any developed countries. The question of how to mitigate and level out the inequality is another, politically-oriented challenge.

Suspension of Imports or Growth?

Some growth critics and environmental debaters claim that there is no such thing as decoupling. The meaning of this is that we are unable to increase global consumption without increasing emissions. According to such critics, this is true even for countries like Sweden and Finland, where the local emissions from traffic, energy production, and industry have been successfully reduced during the course of many years. This is because these countries have simultaneously imported lots of products from Asia and other parts of the world, where most of the electricity is produced from coal. The products we consume in the Nordic countries are burdened with greenhouse gas emissions made in other parts of the world.

From that perspective, it would not be sufficient to reduce the emissions from the activities in our own countries. We would have to reduce our total imports and fundamentally change our lifestyle to achieve an effective decoupling. The conclusion would have to be that not even green technology would be enough to allow for economic growth combined with reduced emissions per capita.

This standpoint is highly questionable.

In the first place, to suspend the import of goods from the rest of the world is completely unthinkable. However, let's make this thought experiment anyway: Finland and Sweden abort the import of consumer electronics and industrial components from Asia and household appliances from Eastern Europe. There is no doubt that the CO_2 emissions per capita would be reduced slightly. However, the effect of this action is negligible on a global level. On the other hand, we would risk being technically surpassed by other countries and eventually being transformed to developing countries when our growth had stalled.

After spending a few years in such a downward spiral, our productivity would have decreased to a much lower level than in the countries which had continued their consumption, economic growth and international trading. As a consequence, our industry and society would no longer have any money to spend on environmental actions. The products that we had earlier manufactured with low CO_2 emissions would no longer be fit for export, because of our reduced competitiveness. This production would also be moved to Asia instead.

The sum of all this is obvious—the total benefit for the climate is reduced.

If instead we continued to modernise our society, business, and industry, we would then produce increasingly better products that support a sustainable development both in our own countries and globally.

The last argument, and the most important one in my opinion, is that by keeping our economy running, we would be supporting techno-evolution. If we developed and exported competitive and efficient products, people in other countries would also be able to use these for the benefit of the environment. Successful products bring benefits in several steps. For example, an electronic device manufactured in Finland could be used to better control electrical motors in cement factories all over the world, thus contributing to the reduction of greenhouse gas emissions in the entire cement industry.

Fortunately, it is becoming increasingly accepted in the political and scientific communities that the climate problem has to be solved through economic growth and technology shifts. There is hope for the planet when decision makers believe in decoupling.

Do the Rich Countries Have an Environmental Debt?

It is a fact that 20% of humanity consumes 80% of the earth's resources. It is also a fact that the industrial nations have dumped increasing volumes of greenhouse gases into the atmosphere for a long time. Many people claim that the rich countries have already exhausted their emission quota and that they are now responsible for the environmental problems. Therefore, it is deemed unfair that the poor countries have to suffer the most.

In the international climate negotiations, the question of a possible environmental debt is a key issue. How should the cost of the climate burden be distributed between the countries? The developing countries rightly claim that up to now, it is mainly the industrial countries that have filled the atmosphere with greenhouse gases; and presently, when the poor countries try to develop their living standards, they must pay for their emissions because of the old sins of the richer countries. The same view is applied to natural resources: the industrial countries have raised their living standards with inexpensive natural resources that have often been acquired from poor countries as a result of colonialism and slavery.

Can we conclude from this that the industrial countries really owe the poor countries an environmental debt?

The answer is not self-evident, because the developing countries' ability to increase their living standards depends on the knowledge and innovations that have been created in the industrial countries. It is, after all, the rich countries that have paved the way for a sustainable development.

The development towards industrialisation started in a very limited part of the world and has proceeded during the course of about 250 years. During that period, the industrialised countries have developed the resources that constitute the foundation for human welfare all over the world, including vaccine depots for almost 100% of the world's population, basic education for more than 90% of the world's children, and electricity for 80% of all people. Additionally, there has been a plethora of useful inventions, like medicines, water purification, hygiene items, and communications equipment. Science, medicine and modern technology appear only in places where knowledge capital and economic resources are available.

Thus, the poor countries win a lot more than they lose thanks to the benefits of the industrial nations' knowledge and technology, despite the long historical period of emissions. They will achieve higher living standards much faster than the developed countries have done, because they won't have to repeat all the mistakes that have been made in the industrialised countries on the long way from poverty to prosperity. One of the most important welfare indicators—decreasing child mortality—shows how rapid the development is in the countries where the modernisation process started later than in industrial countries like Sweden and Finland, Fig. 8.1.

A prerequisite for the development of new technology (medical innovations, solar energy, you name it) is that there are available resources in the form of science, research, stabile communities, entrepreneurship, and financial risk capital. Virtually all the medical and technological utilities that serve us today have been developed in countries where these resources have been available, i.e. the rich countries.

If we take both the plusses and minuses into account, the net result must be that the industrialised countries do not owe the poor countries a debt in this respect. That does not mean that we who live in the developed countries can forget about our responsibilities—on the contrary. We should do our part, not because of a presumed debt, but simply because that we, contrary to the poor countries, have the necessary resources to do it.

The same applies to the statement on an environmental debt for future generations. If all technological development stopped right now, then we would certainly create a debt for future generations.

Our ancestors bathed two times a year, suffered from lice and coughed perpetually in their huts due to the smoke from the wooden fires which was harmful for their lungs. Our generation fares incredibly much better, despite some environmental issues. The next generation will live longer and more healthy lives than we, and their knowledge will be much greater. The debt to future generations will be

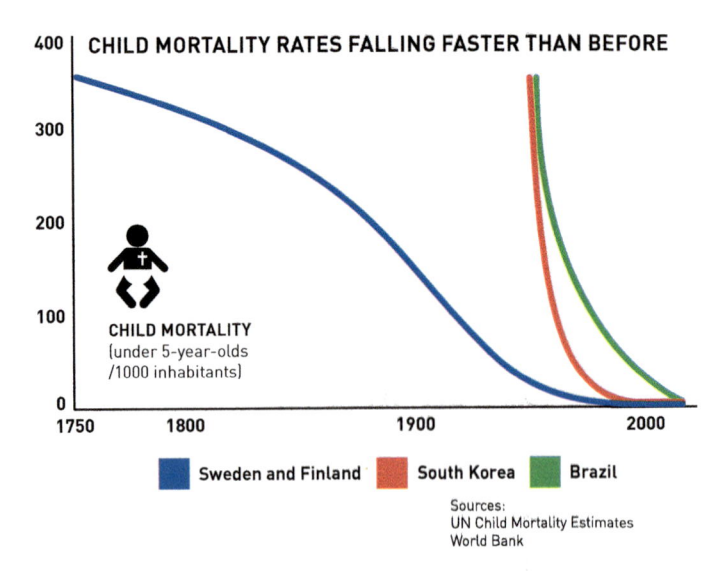

Fig. 8.1 Child mortality is an important welfare indicator. The figure shows that the industrial countries Sweden and Finland needed over two hundred years to reduce their child mortality to its present level. Developing countries like South Korea and Brazil have managed to achieve the same development in just a few decades

worth far less than the heritage they will receive, i.e. all the accumulated knowledge, medical solutions and useful things that have been created up to this point in history.

The Rebound Effect—A Push to the Next Level

However, if the goods and the products are becoming ever cheaper and more energy efficient—wouldn't this simply encourage us, the people in the rich world, to consume even more? If we get less fuel-demanding cars, wouldn't we just drive more often, faster, and longer distances? And if the cost for heating is reduced, wouldn't we just build bigger houses?

Wouldn't the bottom line show increased emissions and resource consumption?

This phenomenon, which is frequently mentioned in the context of a sustainable development, is called the "rebound effect". It illustrates the fact that the more we increase the productivity in our society, the stronger our purchasing power becomes. The effect is that our consumption grows—firstly, each individual can afford to consume more; secondly, more people in the world can afford to consume products and services as prices fall. The consequence is an increased global consumption of natural resources and emissions, even though each single product or service can be produced with less energy and raw materials. That's the rebound effect.

The Rebound Effect—A Push to the Next Level

The rebound effect is a frequent issue in the environmental debate and is often used as an argument against increased growth and consumption, occasionally even as a reason to discontinue the development of cheaper and more efficient products.

The rebound effect exists, and it actually reclaims some of the benefits of increasing efficiency—usually estimated at 5–40%, depending on the consumption pattern. It is also evident that the rebound effect can be even stronger in certain households, even close to 200%. This is a calculation of the consumption increase that will take place when incomes grow. If a family that has lived in very humble circumstances suddenly gets a higher income, the result might be seen in terms of more frequent showering, a house extension, and a new car.

Therefore, some people have suggested that technological development will come to an end unless we start to make it better balanced. There are even political suggestions to the effect that efficiency gains in production shouldn't be used to raise salaries, but to shorten working hours.

Flat screen TVs are frequently used as an example of a technology failing to solve environmental problems. New flat screens are ten times lighter than the old CRT models, which means that they can be produced using just 10% of the natural resources that were once needed for an old TV set. But, as the critics point out, the new TV sets have also become cheaper by the same degree, meaning that many more TVs are bought today than before. This is quite true—thanks to techno-evolution, prices have reduced to the point whereby 80% of world's population now have access to television. Figure 8.2 illustrates the development of TV prices over time.

The weak point in this argument is that once again, the outside world and the technology are treated as if they were static conditions. It is true that the TV sets

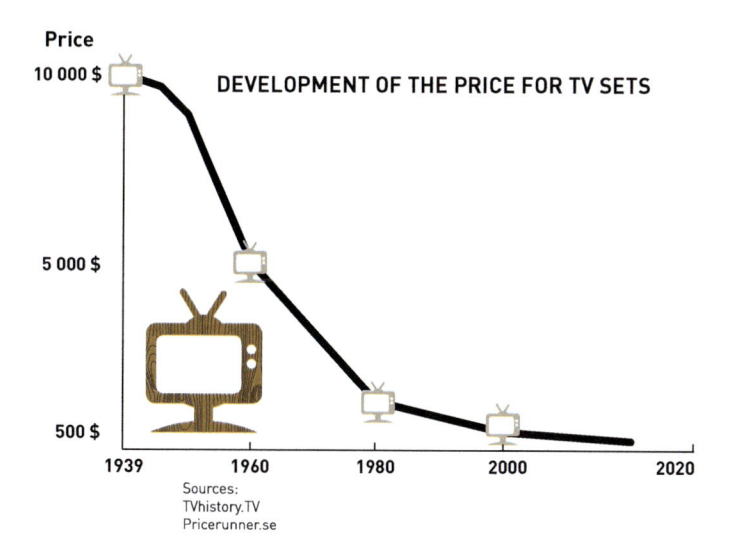

Fig. 8.2 The chart shows the development of the price for TV sets over time. Today, over 80 percent of the world's population have access to television, thanks to technology development and falling prices

have been made increasingly lighter year by year, and that is an ongoing process. Therefore, we should rather see the TV sets of today as intermediate steps on the road to the super thin devices of the future. It is quite plausible that soon we can produce 100 TV sets with the same amount of material that is needed for a single flat screen TV today, and at an even lower price.

However, there is nothing to say that the screens will be much larger. There is a natural limit for the human eye's ability to scan an area. If you have ever tried sitting just two metres in front of a cinema screen, you will understand what I mean. In fact, it is quite possible that the TV sets in the future will be a thousand times smaller and a million times lighter—when direct projection onto eye glasses becomes implemented—a technology that is already under development.

Another concrete example is the rebound effect on cars. The discussion goes that if cars demand less fuel, we will increase our driving and buy bigger cars. Then the gain in fuel consumption and reduced CO_2 emissions would be more or less devoured by the rebound effect.

But is it not more likely that most of us will drive as much as we need, regardless of the fuel consumption? There is evidence in support of this. The Swedish magazine *På väg* ("On the road") recently compared the fuel costs of two car models, the Volvo 240 from 1980 and the Volvo S60 from 2012. It was concluded that it had become 23% less expensive to drive a Volvo over the last 30 years, mostly due to higher engine efficiency. Nevertheless, the trend in the current century is actually towards *less* car driving, according to statistics from Svensk bilprovning (the official Swedish vehicle inspection agency).

The decrease is an average of 1000 km per car between the years 1999 and 2014. The same trend can be observed in Europe as a whole, Australia, Japan, and even the USA. Thus, it seems that no correlation exists between lower fuel costs and more frequent driving. The explanations are many: urbanisation all over the world, car driving just for the experience has lost some of its former attractiveness, and time—more than ever before—is becoming a scarce resource.

Additionally, alternative modes of transport have become more popular than before, and the car has lost some of its value as a status symbol.

In this context, it is worth contemplating the fact that in the 1800s, potential welfare development was considered to be limited by the land area that could be assigned to oat cultivation. Because oats were essential for feeding the horses in a society that still mainly relied on horse power, the oat fields would ultimately reach a limit where further welfare development would be impossible.

The impact of the rebound effect varies a lot between different products and services. But even if a considerable rebound effect exists in some cases, it cannot serve as a general argument against making products more energy efficient. Here are a few examples of how different rebound effects may behave:

- *A factory* that makes an investment in energy-saving manufacturing technology will eventually make a greater profit. If the energy consumption doesn't increase in other parts of the factory because of the profit, the rebound effect is zero. This means that the benefits will be both economic and ecological.

- *A household* changes its old light bulbs indoors to energy efficient lighting but simultaneously expands the lighting outdoors, so that the total electricity consumption remains the same. The rebound effect is 100%, so there is no environmental gain.
- *Car owner A* changes his car, which consumes ten litres of gasoline per 100 km, to one that just requires six litres per 100 km. At the same time, he begins to drive more, so the total annual consumption is reduced by only 20%. The rebound effect is then 50%.
- *Car owner B* changes her gasoline car to an electric car and loads it exclusively with electricity from solar or wind power. The rebound effect is zero—in other words, a good choice from an environmental perspective.

Sustainability Gains Despite the Rebound Effect

During the 21st century, many people have bought new household appliances, either because the old ones have worn out or because they consumed too much electricity. Whatever the reason, we have got new equipment that saves operating costs. A reasonable assumption would be that the saved money has been spent on other consumption, possibly smart phones and e-readers. The debaters who use the rebound effect as an argument for reduced consumption may refer to this fact as an evidence for their thesis: the households consume the entire gain from the more efficient household appliances.

In my opinion, this perspective is far too narrow. The rebound effect is calculated with incorrect boundary conditions. The concept of a rebound effect is not supported by a holistic approach, because the indirect effects are almost totally ignored.

If the money saved by new household appliances was spent on electronic devices, we should of course also take into account all the positive effects created by these devices. When such indirect effects are excluded from the calculation, the estimate of the rebound effect becomes totally wrong. These electronic devices may, for example, contribute to reduced travel, which reduces CO_2 emissions. The e-readers replace physical newspapers and books made of paper, which in turn reduces CO_2 emissions from both the manufacture and distribution of paper products.

In these examples it would be meaningless to speak of a rebound effect. There are indirect positive effects with respect to the environment and the use of resources. These effects are valid locally and—more importantly—globally. If we take the important indirect parameters into account, the rebound effect loses all relevance. Three of these parameters are knowledge transfer, technological evolution, and global distribution of technology.

The rebound effect, when viewed in a longer time perspective (from the mid-1700s to today), is actually nothing other than an improved standard of living over time. In fact, the rebound effect is actually pushing us up to a higher level of living standard.

Chapter 9
Technology Requires Freedom and Responsibility

Abstract Although technology in itself is neutral, the ways in which it is used must sometimes be kept in check. Means of control, whether informative, compulsive, or incentive, are powerful instruments for directing the consumption and technological development in the desired direction. But such means of control should be used with discretion. Since consumption is as important to techno-evolution, it is risky to meddle too much with consumption in general. A forced reduction of certain forms of consumption can, for instance, "preserve" obsolete technology, and result in the prolonged use of resource-squandering products and harmful substances.

Will everything be plain sailing if we simply leave the world's progress to the techno-evolution? No, being a technology optimist doesn't mean that you let technological development manage itself without any human monitoring and intervention. It is not uncommon that the technology used to solve a problem creates a new one instead. Although technology in itself is neutral, the ways in which it is used must sometimes be kept in check.

The list of technological mishaps in the past can be made as long as you like:

Beautiful paints and decorative wallpapers once contained dangerous levels of arsenic that could kill both decorators and residents alike. The luminous digits on clock faces once gave radiation poisoning to the workers painting them. Lead additives in petrol reduced knocking in vehicle engines but harmed children playing on roadsides.

In the early electro-mechanical refrigerators, sulphur dioxide and ammonia were used as coolants—poisonous substances that could be lethal to any users exposed to them. Therefore, the discovery of Freon was regarded as a great step forward—until it was known that Freon was harmful to the ozone layer.

In the building industry there have been many flawed technological solutions and bad choices of material that have created problems both inside and out. And who could know from the start that the amalgam used in fillings—that revolutionised dentistry—would leach mercury into the human body and nature? It took several decades until the results of this "experiment" became revealed. This list can be made much longer, and may easily give the impression that technological innovations invariably create more harm than benefits.

However, all experiments, in nature as well as in technology, are bound to contain a number of mistakes. In a longer perspective, these flaws are unavoidable in the natural trial-and-error process of evolution, where the mutations that aren't fit for life will perish and be replaced by others.

The time this replacement process takes varies from case to case, and ranges from weeks to decades, depending on how fast the problems are discovered and understood. The important thing is that sooner or later, the techno-evolution discards solutions that do not work to our benefit. However, in some cases it may be necessary to hasten the evolutionary process with regulations that protect us and our environment.

The Carrot and the Stick

Means of control, whether informative, compulsive, or incentive, are powerful instruments for directing the consumption and technological development in the desired direction.

As means of control are important weapons in the struggle to protect the environment, it is of interest to take a closer look at them. Prohibitions, taxes, subsidiaries, and carbon credits are examples of efficient ways to affect the evolution of technology.

Prohibitions are the most powerful means of control, as they entirely change the preconditions for a certain technology and even can make it disappear. In some of the examples mentioned above, the use of poisonous substances in paint and gasoline was brought to an end by governmental interference.

A successful international ban against the refrigerant gases used in fridges accelerated a technological change that in turn saved the ozone layer in the earth's atmosphere. According to the latest reports from the UN bureau UNEP and the WMO organisation, the ozone layer is recovering, and scientists seem to agree that the problem will be solved.

But such means of control should be used with discretion. Since consumption is as important to techno-evolution as natural selection is to biological evolution, it is risky to meddle too much with consumption in general. A forced reduction of certain forms of consumption can, for instance, "preserve" obsolete technology, and result in the prolonged use of resource-squandering products and harmful substances.

A thought experiment may serve to show how difficult it can be to find the right means of control to solve an environmental problem. Air travel has been severely questioned by those who regard changing consumption behaviour as a solution for the climate crisis. What would happen in an opposite scenario, with a lot more travelling?

An increased consumption of air travel would of course create an increased demand and production of new planes. Larger production volumes would create more development resources at the plane factories, which would direct the technology changes towards higher fuel efficiency.

The development of aerospace technology is accelerating thanks to increased variation and stronger selection pressure. Thus, in the long run, increasing travel furthers more environmentally-friendly planes.

This is happening right now. At the time of writing, the first planes of the new Airbus A320 Neo model are ready for delivery. This model is a staggering 20% more energy efficient than its predecessor. Airbus expects demand for this product to grow—despite the present low oil prices— because fuel prices are expected to rise again.

Is there a catch here? Wouldn't more planes in the air create an increase of the total amount of CO_2 in the atmosphere, even if the individual planes emit less? Doubtlessly, yes. But can the forces of evolution solve this too?

Yes, they can. If we assume that the conditions of this particular life environment, i.e. the aviation market, are changed so that it will be forbidden, or at least very expensive, to emit CO_2, then techno-evolution will support an elimination of CO_2 emissions. Both planes and their fuels will be adapted to their new life environment.

But if the CO_2 emissions from planes are a problem provoking concern, why not then use more powerful means of control and simply forbid air traffic totally? The answer should be obvious. A sudden prohibition would create a too rapid change that would be fatal for development, like a sudden, overwhelming, and permanent flood. If a permanent flood ever suddenly occurred on our planet, land-living organisms wouldn't manage to adapt fast enough and would simply drown and become extinct. Therefore, moderate means of control that allows enough time for evolutionary intensification would be a better solution.

Too dramatic changes can obliterate an existing technology, which would bring more harm than allowing the technology to change in small steps over a longer time period. This is where reasonable and well-balanced means of control enter the picture.

Many people are suspicious of any means of control supporting renewable energy. This is partly because it is the tax payers who will eventually have to foot the bill, and partly because many mistakes are made in the choice of the means of control. We will probably have to face many blunders, both economic and technological, on the way to a new and sustainable energy system. Tax payers will become frustrated when their tax money is squandered on nothing, and investors will tear their hair out as they watch their capital vanish into thin air.

There is nothing pessimistic about this description. This is simply how things work in an evolutionary perspective: it is through trial-and-error that we reach the next step on the ladder of development. We should remember that our objective is a long-term adjustment to new technological solutions. By introducing economic means of control we change the life environment of the technology, with the result that new solutions are developed at a faster rate. In this context, the means of control should be aimed directly at what is considered to be the fundamental problem, i.e. a redundancy of greenhouse gases in the atmosphere. In that way, the technological diversity may increase and more beneficial innovations will be developed.

At his point, it is too early to tell which investments will bring the most feasible solutions. That is why technological diversity is so important for the transition into a greenhouse-gas neutral energy technology. Note that I avoid the expression transition to a "renewable" energy technology, because that would exclude fossil fuels with CO_2 extraction and nuclear power.

Much has been said about using means of control to *limit air travel*. In my opinion it would be better to look for means of control that would limit *the greenhouse gas emissions of flight technology*.

The CO_2 tax in Norway is a good example. Air travel companies avoid this tax on domestic flights in Norway by using bio fuel. Such incentives support development instead of creating stagnation.

Means of Control and Creative Destruction

Early in the 20th century, the economist Joseph Schumpeter introduced the concept of creative destruction. This means that radically new innovations can destroy and break down entities that are dependent on the old order. Organisations, knowledge and a lot of other things will be defeated and annihilated when new innovations invade their living space.

Technological innovations cause a creative destruction similar to an earthquake. A new innovation corresponds to an epicentre, and the organisations and structures that are wiped out in the creative destruction are similar to the buildings, bridges, and roads ruptured by shifts in the earth's crust. Changes caused by new technology may often seem as painful as childbirth—a new generation is born.

Schumpeter emphasises the entrepreneur as a driving force in this process, but there are also other important elements at work in creative destruction. As mentioned earlier, the inventor's main role is to create ideas (mutations). The innovator plays a similar part, but more in the line of the further development and refinement of ideas (selection and adaptation); i.e. introducing a product onto the market and transforming it into a business. These roles often overlap, and in some cases, the same individual plays both roles.

Here the planned economy of the Soviet Union may serve as an example. There the regime decided which technologies should be developed and which products were to be produced. In other words, the means of control were extremely rigid and powerful. The highest priority was given to the arms technology, space technology and nuclear power. Scientists, inventors and innovators in these fields could get an outlet for their creative energy, occasionally with quite impressive results. But innovations in other industries hardly existed. Neither the computer nor the electric car could ever had been conceived in the Soviet Union.

The main reason for this was that techno-evolution had been suspended because its important driving forces, such as entrepreneurship, were banned. Neither was there any mass consumption of novelties. Without consumption-driven growth, competition and the pursuit of profit, no new products were developed in an efficient way.

The intensity of the techno-evolution is dependent on its environment, and the process will stagnate when its vital elements are amputated.

In my opinion, a political system that empowers techno-evolution is no less important than the regulations and other means of control that tend to flourish on the political agendas of today.

Paralysing Analyses

Presently there are a number of calculation models to compare the effects on the environment with the consumption of resources. The MIPS method, which I have discussed earlier, is one such model. But neither MIPS nor any other similar method take the prolonged time dimension of techno-evolution into account. That is their fatal flaw.

Imagine a scenario where a new invention, a so-called mobile phone, appears as an alternative to the traditional landline telephone. Political decision makers and environmental authorities want to know the facts of ecological sustainability to decide about the relevant means of control and manufacturing authorisation. They decide to make comparative lifecycle analyses of both these telephone alternatives.

The design of the landline telephone isn't very complex; it contains rather few materials. The mobile phone is much more complicated and contains complex materials and unusual metals. Additionally, its production consumes 20 times more natural resources than the landline alternative. (Please remember that the first mobile phones weighed 4–5 kg and were the size of a coffee maker.) The calculation presuppose that the services performed by the products are the same. All you can do with them is make telephone calls, right?

How would a decision maker of today deem this situation?

Governmental decisions based on MIPS or any other similar sustainability analytics model would of course have been in favour of the landline phone. The analysts couldn't have foreseen that the active consumption of mobiles in just a few decades would make them 100 times lighter, 1000 times cheaper and 10,000 times more useful—hard evidence based on actual development. There is no way they could have foreseen how many incredibly useful operations a mobile can manage as soon as it is taken out from your pocket.

The development of the mobile telephone could have been aborted with the help of various means of control, with the justification that it would transgress the limits of the planet because of the consumption of rare metals.

Another hypothetical example would be if our assumed decision makers met to decide about investments in mechanical, handheld calculators or in the first 30 tonne computer ENIAC.

What would the result have been of this comparison, based on a sustainability analysis made in the year 1946?

Today we need only half the land area to produce the same amount of food as 50 years ago. In another 50 years, we will have even better developed crops and cultivation methods plus cultivation in dedicated factories. Sustainability analyses have their uses, but we have to keep in mind that they are lopsided because they do not pay attention to the fact that we live in a dynamic world.

Care and Testing

A common suggestion in the environmental debate is that technological products should be tested and developed more thoroughly in manufacturers' laboratories before being launched on the market. The intension is that we would get better, safer and longer-lasting products.

However, testing products before their market introduction with the intension to avoid *all* possible negative consequences would be next to impossible. Such a procedure would only hamper development and might even delay much-needed, useful products by hundreds of years.

All these phenomena—stability, balance and eased selection pressure—have certain drawbacks in the biological world, and even more so in the world of technology. A too strong focus on the potential risks involved is often a doubtful strategy.

There is no present simulation technology that can predict, for instance, which electronic products we will use in the year 2030.

Proving with absolute certainty that radiation from wireless data communication is *not* harmful would have required long-term studies over a period of at least 30 years. The benefits of improved healthcare, saved lives, increased knowledge and a healthier environment would have had to wait. Omnipotent authorities are frightening in many ways.

On the other hand, there are many examples of technologies from the past with initially undetected side-effect that should have been tested much more thoroughly before the products were allowed to hit the market. In the future, however, a growing arsenal of test and simulation algorithms will hopefully precede direct market selection. In addition, the consumers' selection will partly be transferred to the web, whereby the first selections will be made in the virtual world.

Perhaps computer-based impact assessment systems will eventually be so sophisticated that they can predict almost all possible consequences before a product is introduced on the market. However, until then we must rely on selection pressure and natural consumption selection.

Chapter 10
Resources Are Dwindling—Yet Growing

Abstract If you believe that our civilisation stands and falls with natural resources like petroleum and coal, you consider fossil fuel, and especially cheap petroleum, to be the most essential element in the ascendance of industrialism and modern society. In my opinion, however, fossil fuels are not the main explanation for high living standards in the industrialised countries—rather, the reason lies in brainpower and techno-evolution. One thing that is seldom or never mentioned in the environmental debate is that there are also resources that are constantly growing. These resources are foremost of a kind created by humans themselves, like human health, knowledge, and all products and services that we have created.

With a population of nine billion by the year 2050 and an additional two billion by 2100—all desiring a decent standard of living—it doesn't take much to realise that the planet will have be managed in a smarter way than it is today. And it is not just a question of dealing with climate changes! We also have to deal with the acidification of the oceans as a result of CO_2 emissions, as well as pollution, eutrophication, and other issues.

More worryingly still, it would appear that the raw material resources of the world are becoming depleted in pace with increasing demand. Some leading scientists even claim that soon we will be suffering from a severe shortage of metals and minerals. This concern is understandable if we are to assume that consumption will continue while development stands still.

The concept of natural resources is very wide; after all, we are dealing with about 80 pure elements and more than 5000 different minerals, many of which will fortunately suffice for millions of years to come.

The quantity, recyclability, and resilience of natural resources varies greatly and can be divided into renewable and non-renewable types. The latter are also known as finite resources. The term renewable, as per definition, deals with everything that can be regenerated, for example biomass. However, there are also many exceptions to the definition. Solar energy, for example, is also considered to be a renewable resource. Non-renewable resources are, among others, fossil fuels and minerals. Peat is essentially a biomass, although it is still not regarded as a renewable, like wood is, since its regeneration process is extremely slow. And although metals are not renewable resources, they can be recycled—unlike fossil fuels. There are also other ways to classify and define natural resources.

However, the fact that some finite natural resources will eventually disappear is not an unsolvable problem. One such finite natural resource is petroleum. By the time petroleum has gone altogether, it will have been replaced by solar power and other renewable energy sources. With respect to the environment and greenhouse gases, it would even be beneficial if there was no petroleum left. Whatever the case, the fact remains that petroleum simply won't be needed in the future.

On the other hand, there are critical natural resources that are priceless, like the air, the sea, the food supply and many other things that nature provides, so-called ecosystem services. Forests and fresh water, which belong to the category of renewable resources, are also critical, in that we cannot survive without them.

In this area there is a need for political means of control, in the form of prohibitions or other limiting actions.

However, the concept of sustainable development—as it is commonly defined today—has two important weaknesses. One is that it is very general, so that it allows for widely differing interpretations. The other weakness is that today we cannot know the net value that future generations will find in the things that our generation creates with the natural resources. As such, the values of natural resources and ecosystems should be expressed in concrete numbers—something that is much easier said than done. Moreover, the future value of humankind's efforts is even more difficult to prove.

On the other hand, we have enough knowledge to establish certain frameworks—regarding global warming for example—as was seen at the UN climate change conference in Paris 2015.

The Process Box of Civilisation

To further illuminate the question of what sustainable development really is, we can visualise our civilisation and its planet as a closed box where we enter natural resources and work at one end. At the other end of the box, the modern society emerges, with everything that has been created—including waste, of course. This is the process box of civilisation, which is another way to look at sustainable development.

Maintaining human life, with food, homes, entertainment etc., requires basic raw materials, i.e. various kinds of natural resources. Using these natural resources, we transform our world.

As with all progressive processes, we have to make an assessment as to whether a particular process is value-adding or not. In other words, is the output value higher than the input value, after taking the negative environmental effects into account?

In the past, the process box of civilisation was regarded as being infinite, which was naturally wrong. Nowadays, it is often viewed as a "box of destruction", where the good things in the output are under-valued. On the other hand, emissions and waste are also undervalued, as there is no specified value deduction for some kinds of waste, like greenhouse-gas emissions. Most scientists in this field agree that the transgression of some of the planet's ecological boundaries, like the level of CO_2 in the atmosphere, has caused alarm bells to start ringing. Therefore, the process must be adjusted, but not shut off.

Infinite and Finite Resources

A relevant issue worrying people today concerns what would happen if the input materials (i.e. the natural resources) were eventually finished. Fossil fuels, for example, are some of the most important input materials right now, and are considered to become depleted in the not too-distant future. Many environmentally-engaged people and organisations assert that the industrial world is built on fossil fuels, especially petroleum, and claim that the orgy of consumption will soon be over. The impending fossil fuel shock is even being described as the collapse of civilisation.

However, an increasing number of people—myself included—do not regard the end of the fossil-fuel era as a problem, but rather the opposite. How can such an issue of global proportions be regarded so differently?

It all depends on how great an importance you attach to petroleum and other fossil-fuel sources. If you believe that our civilisation stands and falls with natural resources like petroleum and coal, you consider fossil fuel, and especially cheap petroleum, to be the most essential element in the ascendance of industrialism and modern society.

In my opinion, however, fossil fuels are not the main explanation for high living standards in the industrialised countries—rather, the reason lies in brainpower and techno-evolution. Civilisation has been made possible through innovations—not only in the field of technology, but also in social structure, general science, and culture.

If there hadn't been any petroleum, we would simply have turned to other energy sources instead. In this light, we will have no problems managing without the need for fossil fuels in the future.

The Copper War that Never Started

In the 1970s, it was generally feared that the world's supply of copper would soon be depleted. There were alarming reports in the media that a war for copper was going to break out when China started to extend their telephone network. Today there are far more telephones than could be imagined in 1970; not only in China, but in the whole world. However, no fighting over dwindling copper reserves occurred—on the contrary, copper prices have fallen thanks to the transition to wireless telephone technology.

Today there is concern that rare metals like cobalt, tantalum, neodymium, and germanium will soon become depleted. There is no denying that some metals can become extremely rare, which will make their prices skyrocket as they become increasingly scarce. However, this will simply lead to the development of alternative materials and a reduced demand.

At the same time, and for the same reason, an efficient recycling of such metals will be introduced. The recycling of worn-out products containing such precious metals will suddenly be recognised as a profitable endeavour. This is part of a circular economy, a process that market forces usually manage well on their own. A third consequence is that prospecting and technology regarding the utilisation of metal will be further developed, so that new sources can be found and raw materials will be used in a more efficient way.

Those concerned about depleting resources in the future can rest assured that many of the important materials used today remain virtually unlimited, i.e. silicon, iron, calcium, aluminium, hydrogen etc.—as well as solar power.

Nevertheless, suppose we were to ask ourselves if it could ever occur that a rare metal became totally depleted before a useful substitute was found. In theory, the answer to such a question would have to be yes, although in practice, I do not believe that such a situation would ever occur. It would be utterly astounding that there would have been no innovation made to save the day—founded on principles that would eliminate the very need for the lost resource.

Throughout history, there have been several occasions when a substitute was found as an important resource became depleted. A classic example is the fertiliser saltpetre (or potassium nitrate), a finite resource that was essential to agriculture all over the world a hundred years ago. There had once been great volumes of saltpetre in Chile, but supplies eventually dwindled.

As a resource becomes depleted, the usual turn of events—as mentioned earlier—is that the price for the supply rises, which makes the search for alternatives worthwhile. In the case of saltpetre, a new chemical process was developed to create an industrial fertiliser. Two German chemists—Fritz Haber and Carl Bosch—invented the so-called Haber-Bosch method for the generation of an artificial fertiliser, for which Haber was awarded the Nobel Prize in chemistry 1918. Over time, Chilean saltpetre became fairly unimportant thanks to the artificial fertiliser. Consequently, a natural resource was replaced by an artificial alternative that has since saved millions of people from starvation.

15 years ago, there was great concern that lead (Pb) would soon become a totally depleted metal. Today, when we look at the market prices for lead, there is nothing to support such concern. Lead has been phased out from many products, as the developed countries have replaced it with other metals. In addition, the recycling of lead is very efficient. The same is true for many other important raw materials—cotton being another example.

The Copper War that Never Started

Presently, a large number of alternatives for cotton exist on the market, for example viscose, lyocell, polyester, hemp, and water hyacinth. These textiles have properties that are sometimes even superior to those of cotton. It would be quite possible to replace cotton entirely, if needed. As a matter of fact, many of the new materials affect the environment less than cotton, which requires a lot of water for its cultivation.

Drinking water is a natural resource that is in short supply in many parts of the world. There is, of course, an abundance of water in the world as a whole, but there is a severe lack of clean drinkable water in several places, mainly because of war, distribution problems, and primitive social systems. Thus, clean drinking water is a critical natural resource, albeit it is not a finite resource per se. The problems are of an entirely different kind than the risk that water as a resource would be totally depleted on earth.

Extensive deforestation has consequences for ecosystems in some parts of the world, especially in countries which are not yet industrialised but dependent on an income from expanded farmland and timber. In the USA and Europe, however, the reduction of forest areas was halted many decades ago. Thus, the problems associated with natural resources are different depending on the kind of resources and, even more so, which countries are in focus. Another important fact in this context is that over-exploitation is usually reduced in pace with rising living standards and modernisation.

The pressure on natural resources is eased as new technology is being developed. For example, land-based fish farming can save water environments and emission-free technology can improve the air we breathe. Non-poisonous agricultural chemicals can spare the insects that are needed for the ecosystem services.

The fear that vital natural resources will disappear is old, as it has its origins in the instinct for survival. From that point of view, it is quite natural that people are worried today. This may also be the reason why civilisation critique is frequently rather unfocused. The debaters illustrating the impending end of our natural resources with a number of earth planets, footprints, or "Overshoot Day"—the day of the year when we transgress nature's annual budget—are not always explaining exactly which resources they are including in their calculations.

Up until now, such prophesies of disappearing resources have always been wrong.

Often the availability of a certain resource has been underestimated and new sources have been found. Resources like common metals, fuel, minerals, crops and other biomass materials will probably not be depleted in the foreseeable future.

The second reason why such prophesies have been proven wrong is that ways have been found to extract natural resources from sources that have earlier been deemed as unattainable or unprofitable. The third, and most important, reason is substitution—where new technology makes it possible to create a substitute for the lacking resource, or a new technological solution is implemented that eliminates the need for the rare resource or even its substitute.

We use our knowledge to create new substitutes even for raw materials that are still available in abundance. A few examples are bio fuel as a substitute for petroleum, bioplastics instead of petroleum-based plastics, synthetic rubber, metals replacing other metals, carbon fibres replacing metals and industrial diamonds instead of rare natural ones.

Another current example of a threatened resource is lithium. The demand for lithium, an important component in batteries for smartphones and electric cars, is skyrocketing. However, researchers at Stanford University have recently reported that aluminium—the most common metal on earth—could become a substitute for lithium.

At the time of writing, scientists at Lund University in Sweden have found a way to use iron as a substitute for the expensive and rare metal ruthenium in solar cells. As we all know, iron is cheap and available in large volumes. The business magazine Forum reports that a partnership of university scientists in Helsinki and Barcelona has developed and patented solar cells based on black silicon. One reason for using silicon is that it is abundantly available in nature, like copper and aluminium.

Furthermore, black silicon technology offers a very high thermal efficiency and facilitates the capture of solar radiation at a low angle, which is particularly important in the Nordic countries. And there is nothing to imply that the growing contribution of solar cells to our future energy supply will be limited by a lack of natural resources.

From Waste to Resource

Waste management is an area deserving special attention. The waste problems associated with urbanisation have been a source of poor hygiene and illness from as early as the Middle Ages. In the 1970s, there was a growing concern for the stinking mountains of detritus that threatened to overwhelm cities as consumption increased. The waste mountains were perceived as a menace against both the planet and humanity. The stench and filth were not the only problems—even worse, the leaching of heavy metals and other poisonous substances threatened to pollute the watercourses. At that time, it was difficult to find any feasible countermeasures.

Today, we have a number of workable solutions, and few regard waste as an unsolvable environmental problem, even if there is still room for improvement.

A fundamental waste strategy in the EU is composed of the following steps: minimising or preventing the generation of waste as far as possible, re-using and recycling material, recovering energy from waste, and finally disposal of remaining waste in a safe way. In countries with highly-developed systems for recycling and energy recovery, only 5–10% of waste ends up in landfills.

The waste from farming and industry is also fairly well managed in the developed countries, and the same goes for hazardous waste. There are no longer giant mountains of waste spreading their reek far into our cities.

But the inequality is profound—the waste management policy in poorer countries is still primitive. Nevertheless, the working waste management in the developed countries of today most conspicuously shows that the solution for the waste mountains of the 1970s was not reduced consumption, which many at that time regarded as the only possibility.

The development was exactly the opposite.

As consumption increased the waste mountains shrank. However, the waste management issue has also shown that market forces and technology are not always able to solve the problems on their own. Cooperation between decision makers, environmental authorities and industry organisations is also necessary.

In the Nordic countries of today, metal, paper, plastic, and many other things are recycled and re-emerge as new raw materials. From the remaining waste, energy can be recovered. Some of the organic waste is simply returned to nature, while the rest remains as an intermediate stock resource that will most probably be recycled in the future.

The small fraction of waste that cannot be used in any way is taken care of and destroyed or disposed in a safe way. The part that is still buried in landfills and remains a potential risk to the ecosystem is rapidly decreasing. The overall trend is very promising, in that recycling regulations and the associated technology are being developed while the value of waste material is increasing. I dare to make the prediction that 20 years from now, waste problems will be globally under control—if not finally solved completely.

Growing Resources

One thing that is seldom or never mentioned in the environmental debate is that there are also resources that are constantly growing. These resources are foremost of a kind created by humans themselves. And one of the best examples is *human capital*.

The concept of human capital is generally defined as the sum of human knowledge, ability and capacity. Human capital is continuously increasing on a global level, partly because the population is growing, and partly because increasing numbers of people attend school and receive the benefits of a higher education. The increasing availability of information and learning material is also contributing to human capital.

Human capital is useful for finding solutions to various problems, contemporary as well as future—everything from identifying dangerous meteorites on their way to earth, to new treatments for the sick. It is, for example, an invaluable resource to have talented scientists available when a new virus appears somewhere in the world, or to have experts who are able to manage and solve problems associated with natural disasters. The available knowledge at humankind's disposal in chemistry, medicine, economics and other useful areas is just the tip of an iceberg that represents the enormous potential of human capital. Additionally, there is still a lot of brainpower out there that hasn't had the opportunity to realise its full potential.

In a book I read, a metaphor is given with the intention of illuminating the unsustainable consumption and depletion of natural resources:

> A family that sells a capital asset, their house, can live well for many years on the payment that they received when they sold the house. But when the money is gone, the family has nowhere to live and nothing to live on. The same is true for the resources of nature—and, up to a point, the climate balance.

The writers claim that humankind is consuming the planet's natural resources and that, consequently, we soon won't have anywhere to live. In my opinion, a more correct picture of the planet's state would be that the family sells their house, and uses the money to *rent a flat and improve their education.*

With better education, they can increase their income and buy a new house—or use their new competence to build a better house.

The latter metaphor takes the human capital into account, which the earlier didn't.

A remarkable observation that I have made regarding human capital is that almost every company in the world seems to emphasise that their staff is their most important asset. This claim is repeated time and time again in annual reports, company presentations, staff strategies and recruitment policies.

But in the case of global population growth, I haven't noted a single individual in the business world, in the media, among the politicians or environmental researchers or others, who have had anything positive to say about the fact that we will have an additional four billion people at our disposal at the end of this century.

On the contrary, it is described as a severe problem, a menace against the climate, the natural resources and our ability to support ourselves. Scientists have even designed a mathematic equation that shows exactly how much each new-born child adds to the burden of our already faltering planet:

$I = P \times A \times T$ is a frequently quoted equation in discussions regarding sustainability.

I = environmental impact
P = population
A = affluence
T = technology

This equation was created by American environmental debaters in the 1970s and is still frequently quoted. In the equation, population is only counted as a burden to the environment. As far as I know, environmental scientists and debaters have never regarded population as a positive force, i.e. as an increase of the human capital that may contribute to the reduction of pollution and other related problems.

What if we could regard four billion additional brains as a super resource!

A clear majority of these people will attend school. Most of them will have access to mobile internet connections and digital education up to university level, probably virtually for free. Today there are already free courses published on the internet by American universities. There will be thousands of potential geniuses among these billions. They will be scanned and found by organisations that are looking for innovative and talented people.

So many new brains—connected for problem-solving via the internet—can only serve to increase the human capital of the world in an entirely new and decisive way. A total of 11 billion human brains in interaction with supercomputers, intelligent algorithms and organisational software can, already in this century, create a new mega-resource that may surpass a science fiction writer's wildest dreams. That is a very different way to regard the population increase in the world.

On the other hand, we shouldn't forget that over-population is a current problem in some countries, for example in parts of Africa. In such countries, we cannot simply rely on a modernisation of the society to automatically alleviate abject poverty. It is apparent that several different kinds of aid are needed.

Structural Capital and Big Data

Structural capital is another rapidly growing resource. This concept includes things like system descriptions, documented experience, research results, standards, instructions, models, analyses, directives, statistics, and patents, to mention a few examples. In other words, this is the sum of all the intellectual value that man has created that also exists independently of us. That is how structural capital is described in terms of business economy. Perhaps the world economy in itself, with all its systems, could be included in the concept of structural capital.

Structural capital is different from human capital in that it is the kind of resource that remains in companies and public organisations when the staff has left the building. One example of this is the concept known as *big data*. This can briefly be described as a way to collect and analyse the enormous volume of information stored in digital form, available via the internet. This structural capital includes the data sources, i.e. the information as such, the traces of data traffic and the analytic algorithms. Through sophisticated data analysis, new statistical correlations can be discovered rapidly. Such analyses can reveal patterns which may be very informative, even if there are no records of any cause and effect.

Using big data, researchers can analyse large volumes of internet queries and, for instance, detect an increased frequency of search words that may indicate that a dangerous kind of flu is breaking out in a certain region. That way, we can already discover some impending epidemics faster by big data analysis than through the traditional health organisations. In similar ways, big data can contribute to the mapping of climate changes, criminality, correlations between behaviour and specific diseases, and much more.

There are even experts who believe that big data will help us to predict and mitigate the effects of natural and environmental catastrophes already in the next decade. Some consider this phenomenon to be a new kind of infrastructure that may totally change our world.

Big data is only one example of structural capital and other immaterial resources that will be just as valuable as any natural resource.

Social capital is another resource. This is a way to evaluate the trust and reciprocity that exist between people in a society. There are several definitions of social capital—one of which has been formulated by the Swedish political scientist Bo Rothstein at Oxford University:

> Social capital is the number of people connected multiplied by the level of trust in these connections.

The most interesting thing with this resource, in the context of sustainability, is that it is growing. In practical terms, this means that there are gradually fewer conflicts in the world, and that people are becoming less violent. There is a slow but continuous global peace process in progress, even if we may be inclined to believe the opposite when following the media or when watching the world through a narrow time slot. Although there has been some negative development during the most recent years, the positive trend is clear in the long historical perspective. Whoever doubts this can study the research reports published by World Values Survey. The scientists there suggest that the positive development is supported by industrialisation, trade, modernisation, and economic development.

Their explanation is that we are advancing increasingly further away from societies based on hunting and agriculture, where the control of physical territories was vital to have food on the table. People once had to fight for their hunting-grounds, farmlands, and other natural resources in the vicinity for the sake of survival. Modernisation is gradually eliminating the need for such behaviour. Cultural changes and inclusive democracy are also mentioned among the reasons for a more peaceful world.

International trade and travel are other factors to promote peace. As has been mentioned in earlier chapters, culture, democracy, and trade have always evolved together with technology, which shows that social capital cannot be assessed separately from industrialism.

Artificial Resources Growing at High Speed

Artificial resources are another rapidly expanding resource type. I use this term because conventional terms like "real assets" and "the sum of society's resources" do not quite cover all the kinds of resources that we are dealing with here. Additionally, the concept of "artificial resources" suits well as a counterpart to "natural resources".

This concept includes resources like electricity, housing, vaccine depots, refrigerators, indoor heating, lifesaving equipment, healthcare technology, computers—you name it! In this light, everything that keeps us alive has to be regarded as an artificial resource.

A large part of our material assets—like buildings, bridges, dams and tools—has a very long useful lifespan and will thus be available for many generations. The Caravan Bridge over the River Meles in Turkey is over 3000 years old and still in use today. Thus, we have not only consumed nature's resources, but have created new ones too.

In the near future, artificial intelligence will emerge as one of our most important resources. It will have the potential to solve problems that are seemingly impossible to manage today.

At present, it is true that the natural resources of the world are being relentlessly exploited for every house and electric car that is built. It is also true that the material recycling of scrapped products will solve part of the problem, but not all. Some material is always lost, and some is temporarily tied to the products manufactured.

Nevertheless, artificial intelligence can provide a tool that may solve the entire problem. With the expected escalation of AI, we will gain access to a technological brain capable of managing problems of a complexity that a human brain would stand no chance of solving.

So how will this superintelligence be able to solve the concrete challenges faced due to the consumption of natural resources?

The answer is that it will be handled in several different ways!

In the first place, we will get the means to identify the critical resources and calculate and map the resources available in the world.

Artificial intelligence already provides more reliable weather forecasts. The next AI generation—artificial superintelligence (ASI)—will help us to clarify and manage climate changes in a way that contemporary scientists can only dream of. Thus, we will have a powerful tool that will provide us with more hard facts instead of different levels of probability.

With the aid of artificial intelligence, we may also learn how to use materials more effectively. Today, a common private car weighs about 1500 kg, but there is nothing to say that it couldn't be made to weigh just 500 kg. It's all about optimising the design, the production, and the material used.

Artificial intelligence can also make it possible to develop entirely new materials—e.g. in a revolution of an even greater magnitude than when plastic was invented. Changing from rare to common materials is nothing new, but it can be made much more efficiently with the aid of an increased thinking capacity. Additionally, the sharing economy and logistics can be developed into much higher levels. In the future, we will also be able to develop smarter technology in the field of power distribution. Already today, the supply and demand of electricity can be controlled and better matched thanks to smart technology solutions.

The Essence of Technology—A Natural Resource

Techno-evolutionary capital is a context and an idea that I want to introduce as an additional growing resource. I haven't been able to find an existing definition of this, so I take the liberty of creating one: Techno-evolutionary capital is the value of techno-evolution that occurs without organised human intervention.

Admittedly, the basic form of techno-evolution—as described in the earlier chapters—would not be possible without human beings. However, it would be wrong to treat human capital and techno-evolutionary capital as being exactly the same thing. Techno-evolution doesn't merely exist thanks to human capital—it also transcends this relationship, through basic evolutionary processes that began long ago and are still in force today.

These basic processes are universal and have not been created by humans—they merely follow the same principles and patterns as all other forms of evolution. As mentioned at the start of this book, these evolutionary processes existed on earth a billion years before man—in fact even before any life appeared on the planet.

Big data is an example that shows how a new technological asset emerged into existence, entirely without intent and any human control. No one in the 1990s could have guessed that the internet would offer this opportunity, and no one planned it.

Big data was created by techno-evolution.

It was suddenly discovered by someone—ready to be harvested. In a long time perspective, all artefacts, products, systems and services are subject to this "driverless" development. In one way or another, all technology is affected by techno-evolution.

This is a valuable asset that has lifted us from animals to humans, from cavemen to space travellers; and it is growing in pace with the expansion of the technosphere.

One thing that is obvious is the fact that techno-evolution is working to the benefit of humankind. Right now, for example, it is in the process of creating artificial intelligence and ASI, which will probably prove to be the most important artificial resource thus far in the history of humankind.

In light of the above, techno-evolutionary capital can actually be regarded as a kind of natural resource.

Artificial Resources—A Summary

In conclusion, *the resource increase* is the sum of *human capital, social capital, structural capital, all artificial resources,* and *the techno-evolutionary resource.*

The consultant company McKinsey—a world leader in business and organisational development—has taken a close look at all the revolutionary technologies emerging at this time. In a report from 2013, *Disruptive Technologies: Advances that will Transform Life, Business and the Global Economy*, 12 new technologies are listed. Until the year 2025, the combined value of these economies is considered to represent a staggering total of 33,000 billion US dollars.

Fortunately, the report shows that several of the new technologies will be mostly beneficial to developing countries. As examples, the mobile internet, cloud technology and services, 3D printing and renewable energy can be mentioned. These steps forward will undoubtedly leave good footprints.

In short, the growing resources are: human health and competence, the world's total wealth of science and knowledge, all physical products and services that humans have created, and techno-evolution in itself.

Considering everything that is included in this impressive list, I am convinced that the list should be taken into account when talking about the earth's resources and the global balance sheet.

Thus we have every reason to expect that techno-evolution will ease the pressure on our finite natural resources. In the next chapter, we will discuss our planet's enormous energy potential and how the climate issue may be solved.

Chapter 11
The Climate Issue Can Be Solved

Abstract Among all the resources needed for human survival, energy is as indispensable as water, food and oxygen. Unfortunately, the use of fossil fuels to supply our energy needs has created several problems, of which climate change is the most threatening. The starting conditions for problem solving are excellent. Considering the enormous energy potential, in combination with an accelerating techno-evolution, my estimate is that the energy price in the world may be reduced to a tenth of its present value, during the coming decades. Inexpensive and plentiful energy allows for growth within the ecological boundaries of the planet and can solve most of the current challenges to civilisation.

The use of fossil fuels to supply our energy needs has created several problems, of which climate change is the most threatening. Consequently, we have to find a solution that also reduces the greenhouse-gas emissions in the atmosphere.

At first, this may seem as an impossible task. However, an encouraging fact is that there are virtually endless volumes of clean energy within reach, and we have already begun to harvest them. So there is hope! The only catch is how to bring the new sources of energy into use at a faster rate.

The energy the sun delivers would be enough for 70,000 billion people, if we could use it all. That should be compared to the mere seven billion people living on the earth today. If we could fully salvage just 1% of the sun's total energy input, we would have access to all the energy we will ever need. One percent doesn't sound like much—the potential is enormous!

It is only a question of time before the amount of solar energy we need becomes economically and technically possible to collect. Scientists in this field have suggested that 50 years from now, as much as half of our energy demand may be filled by solar-based solutions. Personally, I think that it can happen much faster—given that there is no breakthrough for some other sustainable and cheap energy source before that.

Energy Supply and Energy System Services

Ecosystem services are something we humans cannot do without; for example, the purification of water and air that is managed by vegetation, pollinating insects, the fungi and bacteria cultures in the soil, and so on.

Ecosystem services are invaluable to us. Likewise, energy system services, such as heating and air conditioning, lighting, food processing, and the power for manufacturing and transportation, are also vital for our survival. Having a reliable energy supply is a prerequisite for a civilisation's existence.

Of all the energy that we use in the world today, 80% originates from fossil fuels, i.e. coal, petroleum, and natural gas, which all produce greenhouse gases. The rest is produced by nuclear power and renewable energy sources. According to the International Energy Agency (IEA), approximately 75% of the energy requirements in 2040 will be fossil-fuel based, despite the development of renewable energy. The main reason for this is due to increased standards of living, especially in the developing countries.

By then, the share of natural gas in the energy mix will have increased strongly at the expense of coal, which will result in reduced CO_2 emissions, since gas is better than coal in this respect.

Nevertheless, to limit the increase of global warming to a maximum of two degrees Celsius, the share of fossil fuels should be reduced further, or be combined with CO_2 neutralising technology.

During the years between 2014 and 2040, the primary energy demand of the world—mainly electricity, heating and vehicle fuel—is calculated to increase by 25%, of which the share of electricity is no less than 65%. Consequently, the production method used for electricity is of decisive importance.

Nuclear power is predicted to double during this period. Solar, wind and hydropower will also increase and are expected to cover 34% of global electricity production in 2040, according to a scenario estimated by the IEA.

The utilisation of wind power production is currently 30% of the installed capacity, while the utilisation of solar power is a mere 20%. This means that the efficiency of solar and wind power installations has to be dramatically increased in order to achieve the stipulated 34% share of the total electricity produced.

If storage technology is developed faster than is expected today, the picture may change. Another method would be to build electrical grids with high voltage direct current, which would make efficient energy distribution over long distances possible. As the wind always blows and the sun always shines somewhere in a sufficiently large geographical area, such a system could level out the varying availability of renewable energy production.

The Research Situation Concerning the Greenhouse Effect

The air layer of the atmosphere becomes thinner the higher we go; 80% of its mass is located under an altitude of 11 km. If the entire volume would be evenly distributed with the same density, the layer would be eight kilometres high.

If we imagine that the earth is the size of a football, the volume of the entire air layer would correspond to that of a table tennis ball. In other worlds, we are travelling through space on a football with an air tank the size of a table tennis ball. The chimneys and exhaust pipes of every car, plane, boat and building in the world are connected to this same air tank. It is like driving a car with the exhaust pipe pointed right back inside.

Common sense tells us that something is bound to happen to our air tank when we burn 1000 barrels of petroleum per second year after year, not to mention the even larger volumes of coal and gas. As we all know, the perceived threats emerging due to the increasing greenhouse effect are many: rising sea levels, storms and hurricanes, water supply problems, ocean acidification, and devastation of the seas' ecosystems.

On the other hand, the earth would be uninhabitable without CO_2 in the atmosphere and the greenhouse effect. The problem is that we are simply getting too much of it. The fact that the greenhouse effect is affecting the climate has been scientifically proven; a relationship clearly exists between increasing greenhouse gases and the warming of the atmosphere.

Carbon dioxide is one of several greenhouse gases preventing the heat radiated from the earth from escaping into space. Methane from ruminating animals is another source of global warming; additionally, ground-level ozone, Freon, and nitrous oxide are included among gases contributing to the same effect. Water vapour also causes a considerable natural greenhouse effect. Of the earth's anthropogenic greenhouse gases, CO_2 accounts for slightly over 80%, methane for around 10% and the rest for less than 10%.

The level of CO_2 in the atmosphere is simple to measure and it is clearly rising. On the other hand, it is not self-evident that the major part of this increase is anthropogenic; that is, created by human activities. However, most scientists agree that this is by far the most likely explanation.

At the same time, there are other phenomena that have a stabilising or cooling effect on the temperature, like aerosols in the atmosphere. The majority of these consist of particles from volcanic activity or other natural sources, but some have been created by human technology, like soot particles, dust and sulphates. Of the so-called anthropogenic aerosols, a considerable share comes from the combustion of fossil fuels and biomass.

There are also natural processes to reduce the CO_2 in the air, for example the oceans' assimilation and the photosynthesis. In the Southern Ocean, so-called phytoplankton assimilate CO_2 and store considerable volumes of carbon. The melting of the icebergs liberates nutrients that fertilise the vegetable plankton, a process which is estimated to store a fifth of the carbon in this part of the ocean.

It is calculated that the planet and its ecosystem currently absorbs 40% of civilisation's yearly CO_2 emissions.

Other systems affecting the climate include solar activity, volcanoes, ocean currents, and the growth of biomass. The climate question is certainly a complicated one, which leaves room for scepticism in the debate. And I, for one, certainly believe that it is important that sceptics are also allowed to vent their opinions; it has happened before that science has got things wrong.

When the climate changes—regardless as to whether the reason is anthropogenic or not—it is wise to start with the climate change as such and adapt to it. Thus, an adaptation to the consequences of the changed climate should be made in parallel with actions to reduce the greenhouse gases. Naturally, steps should be taken to mitigate the problems in regions where the changes are for the worse, due to storms, rising water levels etc. However, there are places on earth that would actually benefit from a rising temperature.

In my opinion, the debate over whether global warming is natural or induced by man is not constructive. We know that greenhouse gases raise the earth's average temperature, and we know without doubt that the CO_2 level has been increasing for a long time. It is our responsibility to do whatever we can to reduce emissions, whatever the reason.

Decoupling from Kaya

Reduced consumption is a very frequent suggestion for a solution to global warming. The debaters supporting this solution are usually referring to a certain mathematical formula, the so-called Kaya equation. This is used to calculate the effects on the environment as a function of population size, welfare, and technology. The Kaya equation is specifically intended to show how greenhouse gases are affected by the growth of civilisation and living standards. The emitted volumes are also affected by energy consumption and the types of energy used.

The equation, illustrated in Fig. 11.1, can be expressed like this:

KAYA EQUATION

Carbon dioxide emissions = Population × Welfare level × Energy used × Emission intensity

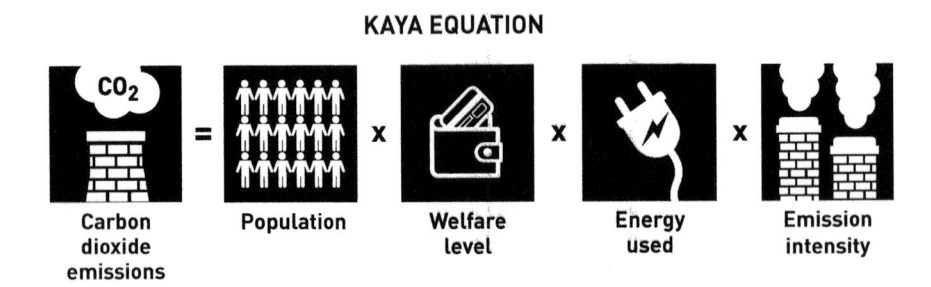

Fig. 11.1 The figure shows that the civilization's CO_2 emissions, according to the Kaya equation, depend on four parameters: the size of the population, the welfare level, the technology used and the emission intensity

*Volume of greenhouse gases = population × welfare × energy intensity
× CO_2 intensity*

Energy intensity is often used as a measure of energy consumption in relation to the GDP. *Carbon dioxide density* in this context is the volume of greenhouse gas emissions in relation to GDP. It is simple to show that the Kaya equation renders a correct result; it is rather obvious, as similar technology is used all over the world and living standards have risen. In other words, we can see that the greenhouse gases have increased in pace with the GDP—these factors have not been decoupled from each other.

Which of the four parameters in the equation can we affect? We know that the first one, the population, according to the forecasts of the UN, will probably level out at 11 billion during this century. Decimating the population is of course impossible, except by supporting family planning.

Some claim that the second parameter, material welfare, can be reduced, at least in the developed countries. The third factor in the equation, energy intensity, and the last one, CO_2 intensity, can be affected by technology changes.

The interesting thing about this equation is that if we set the CO_2 factor to zero, i.e. if we move over to entirely CO_2 free technology, the value of the entire equation becomes zero.

The same thing is true for energy intensity—it doesn't matter how much energy we consume, as long as we use the right technology.

This means that no matter how far living standards rise, the equation will still turn out to be zero. In other words, increased material welfare and continued growth are no problems as long as the greenhouse gases do not increase.

The people who use this equation as an argument for reduced living standards are watching the world in a rear-view mirror. They assume that technological development will come to a halt and that nothing else will change. As we have seen, nothing could be more wrong; techno-evolution is clearly moving towards a point where the last factor in the Kaya equation will be set to zero.

The latest figures from the international energy agency (IEA) for the years 2014, 2015 and 2016 are already indicating that we are approaching a turning point in history, as the increase of the energy-related greenhouse gas concentration has begun to level out while the global economy is growing.

Thus, there are clear signs of decoupling which indicate that global growth can continue without further emission increases.

The idea that decoupling is possible can also be concluded from the fact that the energy consumption per GDP-unit produced was 10% less in the year 2014 than in 2000. International energy experts assert that this beneficial development will continue. According to them, the energy consumption per GDP-unit will be 40% lower in 2040 than in 2014.

According to the world's leading energy specialists, global CO_2 emissions in relation to GDP are estimated to halve during the next 25 years. Decoupling is not only achievable by a transition to new energy sources, it is also attainable through the use of increasingly energy-efficient technology. Although energy consumption will increase by 25% between 2014 and 2040, the energy demand would have been twice as high without techno-evolution.

We are on our way to decoupling ourselves from the perceived threats of the Kaya equation.

Facing Challenges, Seeking Opportunities

There are many challenges to face within the energy issue. First, we still have no way of knowing with any certainty that the CO_2 emissions from our energy consumption really have levelled out and will remain stable in the years to come. Second, there are also other sources for emissions that need to be recognised. Third, even if the increase in greenhouse-gas emissions is levelling out, we still need to reduce the current level of gases already in the atmosphere to avoid dangerous climate changes.

In the short-term perspective, the climate issue may seem overwhelming and frightening. Many feel helpless when facing the tasks at hand. But we should keep in mind that we have more than 80 years to execute the *entire* technology shift that is needed to fulfil the climate goals that have been defined by the PPCC, the climate panel of the United Nations. However, the changes have to show positive results in the year 2020, at the latest, to make it work.

One of the themes of this book is that technological development proceeds exponentially. This means that future technology shifts may be achieved faster, simpler and cheaper than we can imagine today.

There is much evidence for the exponential hypothesis; a recent example is the price development of batteries for electric cars. Just a few years ago, it was commonly assumed that the concept of electric cars was a utopian dream, because of the low capacity of batteries and their high price especially.

Today the picture is totally different. According to a research report from the Stockholm Environment Institute, battery prices have fallen from USD 1000 per kWh in the year 2007 to USD 410 in 2014, and it is expected to fall to USD 160 in 2025. Then it will be close to the limit—USD 150 per kWh—whereby the electric car will be economically profitable in comparison with conventional petrol—and diesel-fuelled cars.

With regard to how techno-evolution works, it is plausible that renewable energy and new products will thoroughly change the electric power supply. One example of such a product is Tesla's Power Wall—an energy-storing battery adapted for buildings.

This kind of product may very well become revolutionary, for several reasons. Firstly, these batteries can save money by providing a small-scale, solar-based electric power supply that saves the environment, and secondly, cheap batteries pave the way for new business models and spinoff products. This would make a lot of difference in the developing countries.

Facing Challenges, Seeking Opportunities 117

The most important element is that individual consumers—not only large corporations—can take initiative and contribute to the power supply, thanks to efficient solar cells and inexpensive batteries. This will accelerate the development in the field. But that is just one piece of the entire scenario. Let us now review some technologies that may lessen or solve the problem of global warming.

There are already a number of promising technological solutions available, even if we disregard the possibility of new inventions which may be made in this century, and which may turn our contemporary technology into historical artefacts. This has happened many times in the past, and therefore it is highly probable that it will happen again. That is not to say that we should calmly wait for a new, hitherto unknown, technology to save us—we have to rely on what we know today.

Energy Production and Technology that Reduce Climate Gases

The technologies presented below are examples of things existing today that can be further developed to make a significant difference to energy production within the next 20 years. These technologies also include solutions that can reduce the existing CO_2 in the atmosphere. Some of the technological solutions presented here have already begun to produce significant results.

- *Renewable energy sources*, that are virtually CO_2 neutral, exist in a number of different varieties. The greatest source is the sun, which can be used to produce either heat or electricity through the use of solar cells. Solar-based power plants will require comparatively small volumes of raw material.

Solar power is commonly regarded as best suited for regions with lots of sunshine and natural heat. But it can also be a good alternative for the northern hemisphere. While the days are admittedly shorter in winter, they are longer in the summer for half of the year. And in the far north, the snow reflects the sunlight to augment the direct solar radiation.

Presently, solar-based energy supplies are growing faster than the alternatives. According to the IEA's estimate at the time of writing, solar-based energy will account for 27% of the world's energy production in the year 2050. Of this energy, 16% will be produced by solar cells and 11% by thermal solar power plants.

Nevertheless, significant challenges still remain which need to be solved. The effect per area unit has to be increased, the price must be reduced further and the energy storage capacity has to improve. But the basic conditions for solar energy, regarded as a technical process, are very good. After all, the fuel is entirely for free and is delivered straight to the production site, contrary to fossil fuels that have to be refined and transported in several stages.

Wind power is already an established technology that is still being developed and expanded. The power per wind turbine was low in the beginning (0.4 MW in the 1980s), and was something that was often frowned upon. At the time of writing, the top power is 8 MW per wind turbine. Researchers at the energy department's laboratory SNL in the USA are developing a new type of rotor blade that is expected to provide no less than 50 MW per wind turbine.

Sea power from waves, tides, streams and salt gradients are also significant sources of energy, not to mention the hydropower from conventional run-of-the-river power plants.

The group of renewable energy sources also includes biogas, which is produced by the anaerobic digestion of biological material such as food left-overs, manure and agricultural waste. Biomass, like wood, tree and paper waste, has a significant and far from fully exploited potential that corresponds to approximately 20% of the world's current energy demand.

- *Natural gas* has shown its potential for reducing CO_2 emissions in recent years, especially in the USA.

Germany and the USA have completely different strategies for their energy supplies. During the years between 2005 and 2012, the USA managed to reduce their CO_2 emissions 3.5 times more than Germany. The reason is to be found in the energy mixes used. The USA switched to using more natural gas while installing only a relatively small share of solar and wind power. Germany, meanwhile, began phasing out nuclear power and expanding their share of solar and wind power, although they have been forced to still use coal.

Thus, the USA has reduced its dependence on coal but strongly increased its share of natural gas. In the short-term perspective, the American way seems to be preferable from an environmental point of view. However, no one knows how the development will turn out in the long term.

- *Fossil fuels and CCS technology* means that greenhouse gas emissions from fossil fuels like coal, petroleum, natural gas, plastic waste and peat can be considerably reduced. CCS is short for Carbon Capture and Storage.

This is a process, mainly based on cleansing with ammonium, where the CO_2 is removed from the emission gases and returned to its source, located far down in the earth. The ammonium can be salvaged and recycled. CCS technology is by no means new; it has been thoroughly tested on a large scale, for example with the exploitation of natural gas in Norway. There, the CO_2 is separated from the fuel gas and pumped back down into the gas fields.

If CCS technology is used for power plants running on biomass, biogas, and waste, the reduction of CO_2 in the atmosphere will be greater than if nature was left to process these substances. This system is called Bio Energy with Carbon Capture and Storage, BECCS, and is regarded by the UN Panel on Climate to be an important method for reaching the emission goal.

Some people believe that CO_2 is just a harmful waste product, but that is not the case. Plants are not the only things dependent on CO_2 in the air. There are also several industries utilising CO_2 as a raw material, a technology that is known as CCU (Carbon Capture and Utilisation).

One of the biggest problems when extracting CO_2 from emissions or the air is that a lot of energy is required, and if that energy is produced with fossil fuel, there is not much to gain.

But on Iceland, for example, where there is a plethora of renewable energy sources, low efficiency is not an obstacle for this method. In Reykjavik, the excellent fuel methanol is produced for export from CO_2, thanks to a surplus of geothermal energy from the earth crust. As methanol is useful for storing energy produced by solar cells and wind turbines, we can probably expect further technological development of methanol production based on CO_2 in the future.

In Tokyo, considerable volumes of the extremely durable plastic product known as polycarbonate are produced from CO_2. In Finland, the Aalto University and Åbo Akademi University—in cooperation with the steel company Rautaruukki Oyj—have developed and patented a method for transforming worthless slag from steel manufacturing to the useful substance calcium carbonate by consuming CO_2.

This method is energy efficient, and it consumes CO_2 at the same time as the finite natural resource calcium is recycled. The first pilot facility in the world is up and running, and in a few years the first demo facility is expected to be built at the Rautaruukki ironworks. The steel industry is currently responsible for 6–7% of the world's CO_2 emissions, and the potential reduction is high.

Scientists at Stockholm University have developed a material with the ability to bind CO_2 from the air, which earlier has been difficult because of the water that also resides in the atmosphere. This material, which is based on copper silicate, can after further development be used to reduce CO_2 emissions in the future. The carbon that is separated in the process can be used in processes for producing other materials.

The technological development exemplified by these two Nordic examples is still in its infancy, but it shows that there are opportunities, not only to remove an undesired substance but to use it to our benefit.

Accordingly, CO_2 can be reclaimed from exhaust emissions or the atmosphere and used as a raw material in various ways. According to an article in the science magazine *Nature*, it will be possible in 20 years from now to produce methanol fuel that will reduce the world's CO_2 emissions by at least 10%, when the corresponding volume of fossil fuels are no longer needed. Together with all other methods to use CO_2 from the air as a raw material, this can be one of the most important weapons in the struggle against the greenhouse effect.

At the time of writing, CCS and CO_2 recycling are not yet economically profitable solutions, except in exceptional cases. According to Exxon Mobile, the cost for CO_2 reduction by CCS is in the same price class as wind power but a cheaper alternative than solar power.

A challenge in this context is that while solar and wind power can be tested on a small scale, CCS demands a much greater effort, and a full-scale facility—that has to be constructed at a cost of millions of euros—might unfortunately still turn out to be a dead end.

At the time of writing, there are 15 full-scale CSS facilities in operation in the world, and seven more are under construction.

Considering this particular technology track, it would be a doubtful policy for a country to advocate a "fossil-free nation" as a political objective. With such a policy, the development of CSS technology would be discontinued, which would be unfortunate with regards to the enormous volumes of fossil fuels that are still available. Be as it may, these fuels will be in use during the whole of this century, and the infrastructure for their consumption is already in place. Additionally, it would exclude the interesting option to remove CO_2 from bio power plants, a climate effort that could hardly be overrated.

In my opinion, the overall political energy objective should rather be "a carbon dioxide free nation". That is a goal that all the nations of the world should embrace, as the climate in the Nordic countries is affected as much by emissions made in Asia as by the emissions from Stockholm and Helsinki.

Reduction of existing CO_2 in the atmosphere is also an interesting option that can be accomplished in different ways. One of the most obvious ways to achieve this is by planting forests. Trees and plants assimilate CO_2 during their entire lifespan and function as a storage place for coal. Saving the remaining rainforests of the world is obviously a smart move. If the net volume of the world's forests could be sustained, it would profoundly affect the CO_2 level in the atmosphere, because forest mass and other land vegetation absorb nearly 30% of all man-generated CO_2 emissions (according to a NASA study).

In addition to the examples mentioned above, there are several other technologies with a very high potential for providing energy and reducing greenhouse gases. Two of these technologies are geothermal energy and natural photosynthesis. You can read more about these technologies in the appendix "More Future Power" at the back of this book.

It is not my intention to present a comprehensive account of all the solutions presently known; rather, I want to emphasise the diversity and the great number of options available. Some of the described technologies have such a high potential that just *one of them* might be sufficient to solve the imminent problem of the greenhouse gases our civilisation has created.

The controversial nuclear power, that provides basic energy regardless of the weather conditions, has also a potential to lessen the CO_2 problem. However, this is subject to the conditions that the construction of the power plant doesn't cause too high CO_2 emissions, and that it will be operative for at least 50 years. The oldest nuclear power plant in Finland is located in Lovisa. It was built in 1977 and no decommissioning plan is known at the time of writing. The oldest operative nuclear plant in the world is to be found in Doettingen in Switzerland. It was started in 1969 and is still producing electricity in 2016.

The economic competitiveness of nuclear power is questionable, and for that reason it is not considered to be feasible in all countries. Other issues under discussion focus on whether the new generation of nuclear plants is safe against major accidents which may result in a meltdown, or if the waste can be disposed of in a reliable way. Be as it may, nuclear power has undeniable advantages and is an important option for CO_2 reduction in a global perspective. Let us, for example, presume that the greenhouse gas problem will prove to be even more serious than so far believed. Perhaps then, the new generation of nuclear power plants may be a feasible solution.

Geothermal energy is available in volumes that are more than sufficient for human needs. And, the wind power potential in the world is also enormous.

Various technologies to remove CO_2 from processes based on fossil fuels are being developed. Therefore, it would be unwise to exclude fossil fuels from the energy mix. As far as fossil fuels can be used without creating greenhouse gases they should be considered acceptable, at least until they become economically uncompetitive. Still, it is plausible that fossil fuels like petroleum and coal will be discredited as unprofitable in the future.

In the next 20 years, energy production and CO_2 reduction will probably be based on a combination of the technologies mentioned above, and some additional alternatives may appear. Exactly which technology will dominate in the long run, however, is difficult to predict. This depends on many aspects—technological, economic, political, emotional, and security related. Moreover, as always, techno-evolution is difficult to predict, meaning that technological diversity is good.

Increasing Savings Demands

The cleanest and most plentiful "fuel" at our disposal during the next two decades will be found in the savings made from increased energy efficiency. In this regard, a broad spectrum of technological solutions is available to optimise our methods of energy use. According to the International Electrotechnical Commission (IEC), there is a global efficiency potential of no less than 30%, and this only includes the technologies that we know of today. This efficiency potential is found in the four great energy consuming sectors: industry, transportation, energy production, and construction.

Some examples of the possibilities of increased energy efficiency:

- *The winds of energy efficiency are blowing*—All energy consuming products will be made more energy efficient. This is not just wishful thinking but a natural continuation of techno-evolution, a process that has been going on since the dawn of industrialism. This has a great impact in a global perspective. When a billion households replace their old energy-consuming appliances with new ones, we are speaking of energy savings corresponding to hundreds of big power plants.
- *The transport business is moving towards climate neutrality*—Cars, planes, and ships can become more fuel efficient—a development that has been in process for a long time. Only in the last decade, the fuel consumption per kilometre has been reduced by tens of percent. It seems probable that energy efficient cars, and specifically hybrid cars, will take over the market in many countries.

- *Increased energy efficiency* on a large scale can be expected when LED technology takes over the scene in earnest, when our electric power grids become smarter, and when "the internet of things" makes its big breakthrough. Read more about that in the appendix "More Future Power" at the end of this book.

While on the subject of energy efficiency, we must keep in mind that on a global perspective, it isn't enough simply to achieve sustainable CO_2 emissions—we also have to allow for a strongly increasing energy demand due to population growth and welfare distribution.

Other system shifts, in combination with behavioural changes, will also lessen the energy demand and reduce emissions, despite sustained or even improved performance.

Super Technology

The technologies we have discussed so far are either already in operation or expected to become operative in the next 20 years. However, there are also solutions that are known but will require a lot of research and development until they are ready for implementation. In the best cases, some of these new technologies may be introduced in the technosphere within a couple of decades from now, but in all probability it will take considerably longer, perhaps to around the year 2070. Although that may seem far in the future, it may be the point in time where the need for CO_2 reduction reaches its real critical peak.

- *Fusion power* is the energy that powers all those billions of stars shining in the universe—a kind of nuclear reactor where atomic nuclei are fused together, instead of split as in conventional nuclear power plants.

Even if nuclear fusion has been achieved in small-scale research reactors, its implementation in commercial reactors is still quite distant. It is a common ironic saying that fusion power lies 30 years in the future and will always do so. Anyway, it seems improbable that this kind of power will be commercially operative within the next 20 years.

The objective of the ongoing fusion research is to find a way to produce a lot of energy without any harmful greenhouse-gas emissions or radiation. In practice, fusion power is created by fusing atomic nuclei of the hydrogen isotopes deuterium and tritium, which can be extracted from common seawater with small volumes of lithium.

"There is no pollution. There are no weapons applications. The fuel is available to all nations," says Miklos Porkolab, manager of the Plasma Science and Fusion Centre at the Massachusetts Institute of Technology. Hopefully, the large-scale fusion reactor ITER in France will be able to deliver power to the electric grid in 2040. But even if it will be delayed, this accomplishment will have the potential to be the greatest achievement in the history of civilisation. The advantages of fusion power are many, and the fuel is just small volumes of seawater.

Additionally, the process cannot go haywire like conventional nuclear power—the process will simply stop if a system fault occurs; the problem is rather to keep the process going. The bi-product is helium, a common and harmless noble gas. Here, we can glimpse the possibility of a future world with a virtually endless energy supply, almost for free. It is not implausible that the young children of today will live to experience it.

Some other promising future possibilities are artificial photosynthesis, super-conductivity, and fuel cells. These are briefly described in the appendix "More Future Power".

The Ace up the Sleeve

In addition to the technologies accounted for, entirely new discoveries are always possible, perhaps already in the current century. This has happened many times before in history, and the possibility of such a "Technology X" is greater than ever. Again, we cannot rely on such a breakthrough to save us; we have to base our actions on what we know today.

My intention with these examples of different energy options is to make clear that the problem with greenhouse gases and other planetary boundaries are not caused by a lack of technological solutions. There is much to indicate that neither is cost a problem, at least not in a longer time perspective.

In fact, the more different types of power plants and energy-saving solutions that are developed and constructed, the better. Today, no one knows for sure what the actual future demands will be, or which technological methods will be successful and which will prove to be dead ends. As noted earlier, technological diversity is just as important as biological diversity.

Energy is available in virtually endless volumes, and we have an entire arsenal of technologies that can stop the increase of greenhouse gases and lessen the planet's burden.

The starting conditions for problem solving are excellent. Considering the enormous energy potential, in combination with an accelerating techno-evolution, my estimate is that the energy price in the world may be reduced to a tenth of its present value before the end of this century. And the price rally will continue.

Low energy costs are beneficial also for other reasons than those which most readily come to mind. With cheap energy, both seawater and polluted water can be made drinkable. Today, that method isn't economically feasible except in regions with a surplus of energy. But the technology exists and has been used for decades on large ships, where there is much redundant energy from the engines.

When cheap, emission-free energy becomes generally available, clean water can be deleted from the list of critical finite natural resources.

Inexpensive and plentiful energy allows for growth within the ecological boundaries of the planet and can solve most of the current challenges to civilisation.

And, as has been shown above—there is no lack of options.

Chapter 12
Scenarios for Success

Abstract Although a global greenhouse gas reduction agreement exists, living up to it will be an even greater challenge than it was to create it. However, postponing the actions required to reduce greenhouse gases would be perilous, since there is always a risk that climate changes will accelerate at a higher rate than expected. Trying to predict the future is always risky, but that's a risk I am willing to take. I claim that the increasing greenhouse effect is not a question of destiny, there are already a great number of promising solutions and signs.

I once visited a trade show in the USA at the beginning of the 1990's. The focus of the event was waste management and my intention was to market the biogas food-waste processing technology that our company had developed. With this technology, methane gas is produced from household waste for use as indoor heating or car fuel. As the waste is processed in a closed system, greenhouse gas emissions from landfills are avoided.

At the trade show, I approached the CEO of one of the largest waste processing companies in the country. He was a great man, also in the physical sense. I told him briefly about our new technology and its exceptional advantages. He looked down at me absent-mindedly and said in a deep grizzly bear-like voice: "Not interested!"

My usual tactic was to counter with: "But, you know …" and make another attempt, but not this time. The time was clearly not ripe to speak about greenhouse gases, and definitely not in the USA.

Since then, much has happened. The greenhouse effect and global warming have been widely recognised in the media all over the world. The governments of many countries have engaged themselves in this issue, and no less than 195 nations have signed a treaty of actions to counter the greenhouse effect.

The energy supply of the world is undoubtedly the leading cause of the greenhouse effect created by humans. As scientists assert that humankind is responsible for over 80% of the problem, the question concerning how we produce and save energy is one of crucial importance.

The Politics of Climate Challenges

Global warming is by far the dominating environmental issue, and the big challenge is of course how the nations of the world will be able to realise the necessary reduction of greenhouse gas emissions. Although a global agreement exists, living up to it will be an even greater challenge than it was to create it. It is quite obvious that well-designed and well-adapted means of control would be desirable to get the necessary technologies up and running.

The scientists of the UN's Climate Panel (IPCC) assert that global warming should be limited to less than 2 °C, preferably 1.5 °C, compared with the pre-industrial level at the beginning of the eighteenth century.

To achieve that target, global emissions should begin to decrease in 2020 at the latest, and be halved in 2050, compared to the level of 1990. Additionally, the emissions should be reduced to almost zero at the end of this century.

The climate scientists of the world—spearheaded by the IPCC—have involved virtually all of the world's governments in actions against climate change. Decades have been spent on discussions as to whether the agreements should be legally binding or optional, whether some countries should be allowed to release larger volumes of greenhouse gases than others, and about the funding.

The various political solutions of global warming can be grouped into a number of different scenarios:

Scenario 1: *Political global agreements that are unequivocal, legally binding, and embraced by all the nations of the world regarding CO_2 emissions. They are unconditional by nature.*

No legally binding agreement was signed in Paris 2015, but more climate conferences are in the pipeline. Strictly binding agreements would stimulate development of the technologies mentioned above. In this respect, it would not be difficult to reduce the CO_2 emissions from a purely technological perspective, given the existing calculations of what is needed to reach the target of 2 °C. But investments in new solutions naturally requires financing, which will temporarily reduce the funding available for other social development.

In order to achieve the climate goal of a temperature increase of 2 °C maximum, emissions have to be reduced roughly 60% in the next 40 years.

The three most critical areas to be dealt with are the production of heat and electricity, transportation, and the general energy efficiency of all energy-consuming products and activities.

As already noted, it is technically possible to fulfil the climate objectives. However, from an economical perspective, the situation is far more complicated. Although the global economy as a whole would hardly be affected by the actions needed to achieve the targets, there are difficult political implications that will need to be addressed as a result of big differences in wealth and democracy between the world's nations.

Of the other emission sources that are not included among the three mentioned above, agriculture accounts for approximately 14%, landscape changes (like deforestation) 12%, industrial emissions not related to energy production 4.3%, and waste 3%. Of these, waste management is the technically simplest problem to solve, which means that 3% of CO_2 emissions can be eliminated rather easily.

But even in this case, there are two conditions which have to be fulfilled. For a nation to even consider prioritising climate issues, it must have a reasonably sound economy. In addition, some means of control have to be introduced for modernisation and organisation of actions.

The emission from industrial processes can be reduced by half during the next decades, despite increased industrial volume. This estimate is based on how much the emissions per item produced have already been reduced in a historical view, plus the state of available technology.

The conclusion must be that what really matters is to focus on electricity and heat production, transportation, and the general efficiency of energy-consuming activities. As the anthropogenic greenhouse gases mainly originate from our energy consumption, a transition to CO_2-free energy would be sufficient in itself to fulfil the stipulated climate targets.

As stated in earlier chapters, efforts to reduce general consumption are to no avail. What is really important, however, are various incitements—directed to consumers, institutions, and companies—which encourage investments in systems and products dedicated to reducing the emissions.

If energy becomes more expensive, the cost will be fully transferred to the products and services we consume. However, according to many calculations, global GDP growth will only be marginally reduced. The Swedish scientist Christian Azar has expressed this fact: instead of being ten times wealthier than today in the year of 2100, the world will have to wait another two years for this to happen.

No one can calculate future changes of this magnitude exactly, but if the economic impact really is that marginal, perhaps people will accept the expense, especially if it is backed by international agreements. At least, the acceptance should be considerably higher than with moral lectures advocating a changed lifestyle.

Scenario 2: *There are political agreements on global regulations; but in reality, they are interpreted as guidelines rather than binding commitments. Does this mean that they are powerless?*

If global climate agreements in reality are interpreted as nothing more than guidelines, they may still be more powerful than one might expect.

Some critics are sceptical of the Paris climate agreement and suggest that the commitments that have been legally determined are of such small importance for global warming as to become meaningless. They may actually correspond to a reduction of a few tenths of a degree at the end of this century.

The fact is, however, that improvements have already been going on for a while, before any agreements were made.

There are strong indications that a change actually began many years before the Paris conference; in a case where reality ran ahead of politics.

2014 was a turning point in history, since it was the year when the increase of greenhouse gas emissions due to human energy consumption levelled out, and we still had global economic growth. This was the so-called decoupling effect that proved that we can increase consumption in the world without increasing emissions.

Such good news is due to the fact that energy-saving technology is gaining ground in the world. Additionally, newly built energy production based on renewable energy sources increased faster than fossil-based facilities.

How was that possible, even before the global climate agreement was in force?

One explanation, which has nothing to do with agreements, is techno-evolution.

Industry always tries to manage with a minimum of energy consumption and raw material resources because it saves cost, which is a strong driver of technological development.

Private individuals are also unwilling to squander energy and are becoming increasingly aware of the environmental problems. Accordingly, technological development is being driven by an increase in the general interest towards energy-saving and climate-friendly technology.

In recent years, considerable breakthroughs have been accomplished in several technological areas. As a result, traditional fossil fuels like coal and petroleum are becoming affected by increasing competition.

However, the climate agreement is also of great importance. I believe that it means a lot more than the marginal improvements that the critics frown upon. In business and industry, there is a growing awareness of stricter emission controls and the rising demand for CO_2-neutral technology. It is simply an opportunity that the business world cannot afford to ignore.

As we know, the market—and the stock market in particular—reacts neither in real time nor exclusively based on facts, but from what its agents believe will happen in the future. Or rather, what they believe that others believe will happen. No one wants to miss the boat, especially when there is an opportunity to share in a growing new business. Investors in the entire industrial chain will look for growth in areas where there is money to be made from environmentally-friendly technology.

Therefore, not only is there an impact from existing emission agreements—the anticipation of future agreements works fine too.

The Paris agreement will be followed up by a new conference in five years. The objective then will be to define stricter demands, which is something that investors and companies are already planning for in their strategies.

Thus, technological shifts are occurring even without agreements, albeit slower. The climate agreement is simply speeding up a natural development.

There are many indications to support the fact that the technological transition has already started. A change of trend was already visible in 2012—around 70% of new power plants built in the EU were based on renewable fuels, according to the Renewable Energy Policy Network for the 21th Century, REN 21. Additionally, energy efficient systems and technologies were in more demand than ever before, despite the absence of global climate agreements. The greenhouse gas problem will probably be solved even if Scenario 2 occurs.

Scenario 3: *It may turn out that global warming is accelerating faster than the majority of scientists believed, i.e. according to the most pessimistic assumptions. In this case, we are facing a higher warming of the earth's average temperature by several degrees, even if political climate agreements are fulfilled.*

Because the warming process is so slow compared to the average human lifespan—and since it doesn't affect all the regions on the planet at once—it may be difficult to stop the emissions fast enough in this scenario. In the worst case, the warming process will not be halted even if we stop all dumping of CO_2 into the atmosphere. In such a critical situation, we must resort to more dramatic and dangerous actions.

If global warming escalates, geoengineering would be a fast remedy for the "fever" as such. Naturally, this would mainly be a method to relieve the symptoms. This can be accomplished using various methods to reduce the solar warming of the earth's surface. The real cause of the illness, CO_2 emissions, is of course what should have been acted upon in the first place; but in this situation, it will be necessary to buy time.

This may not become apparent until a decade or so into the future, by which time science and technology will have been further developed from today's standards, which will facilitate the medication. Here I have to return to the subjects of techno-evolution and knowledge expansion.

If we assume that the problem will have to be solved 50 years from now, we will probably have artificial resources at our disposal that we can't even imagine today. A lot of things are going to happen in science and technology during the next 50 years, probably more than has appeared in the last two centuries.

An entirely different scope of knowledge and technology will then be available.

Another important reason to continue the research in geoengineering is that entirely natural phenomena, for which man is not to blame, will sooner or later threaten our planet. Then, if not before, geoengineering may prove to be the salvation of humankind. The development of geoengineering can be regarded as a life insurance—to be used as a last resort—and not as a first-hand solution for anthropogenic climate changes.

Technology Shift Now or Later?

An interesting question concerns whether the countries of the world should invest in—and pay for—a technology switch-over now or later. It is a question of risk calculation. If we were 100% certain that the polar icecaps are melting, that the oceans will rise several metres, that the ocean currents will change direction, and that a new ice age will come, then powerful and immediate decisions would be simpler to make. But today no one knows with absolute certainty when it is going to happen and exactly how dramatic the effects would be. A rapid transition would cost a lot, at least in the short run. Also, we know for sure that at this moment, millions of people die for other reasons than global warming.

Each year, several million people die from smoke poisoning as a result of burning sticks and dung on indoor fires due to an absence of electricity. Bad indoor air—due to primitive fuels and fireplaces—kills more people than any other kind of pollution.

Consequently, scientists at the Copenhagen Consensus Center, among others, suggest that if more money could be directed to ensure that the 1.3 billion people currently lacking electricity could have access to it, modernisation and economic growth would be achieved much faster in their countries. As a result, many environmental problems would be reduced and millions of lives could be spared.

Malnutrition, malaria and the lack of fresh water are other examples of how the want of basic modern amenities destroys millions of lives. At the same time, billions of dollars are spent on renewable energy. Which priorities should be governing the distribution of available funds? The research team mentioned above claims that right now, there are even more urgent issues to resolve than a reduction of greenhouse gases. Even so, they agree that some investment should be made on technological research to pave the way for climate-friendly solutions in the future.

In my opinion, we cannot ignore the IPCC and the scientists claiming that something should be done against the climate changes right now. Postponing the actions required to reduce greenhouse gases would be perilous, since there is always a risk that climate changes will accelerate at a higher rate than expected. In addition, climate changes affecting prosperity will have a double impact on the poor countries, as the result will be reduced harvests and other problems in these regions. If the industrial countries are also affected, their capacity and willingness to help the developing countries may be even less than before.

It is like facing an emergency on a plane—the adults should put on their oxygen masks first, then assist the children with theirs. If the strong ones are struck down, the weak will become even weaker. Therefore, something has to be done now. Postponing investments in renewable energy production while waiting for research and laboratory work to reveal perfect solutions is simply not good enough!

I claim that competing full-scale and commercially valid projects are just as important as scientific research, with or without subventions. In my experience, such business ventures—in parallel with ongoing research—often have the power to accelerate technological development more than anything else.

It is out in the field, far from the laboratories, where presently known technology has the greatest possibilities to find new applications. For example, using building materials as solar collectors, integrating solar cell technology into building elements, or combining different energy sources.

Other examples are the development of small-scale electricity production and the sharing economy. Electricity producers can deliver their redundant power from their solar cells and share it with other small-scale producers. This is a new trade system that is developing thanks to ever more efficient broadband connections and digitalisation.

Intermittent electric power, —like solar, wind and hydro energy, is difficult to predict, since it varies a lot depending on the weather conditions and is thus unreliable from a distribution perspective. But that problem can be mitigated by better energy storage, more efficient power distribution, and a good combination of power plants providing basic and balancing power.

However, such complementary systems and storage technology will not be developed if there is no intermittent electricity in the grid. Thus, it is a chicken-and-egg situation, and another reason to start expanding solar, wind, and hydro power instead of waiting for results from scientific institutions.

In my opinion, all existing kinds of energy and power solutions should be allowed to develop. As techno-evolution cannot be fully predicted, all possible solutions should be given a chance to flourish. That is not to say that all kinds of power production would be suited to all countries, however, because the conditions are varied.

For my own part, I hope that Scenario 2, which is presently known as the Paris agreement, will work out.

If political objectives, market forces and techno-evolution are not sufficient, the agreement has to be made more legally binding in the future. Then Scenario 1 must be realised. However, I hope that the world leaders' insights have already grown strong enough to make the nations of the world live up to the objectives. I also hope that they realise that the correct diagnosis for treating global warming is neither overconsumption nor a basically incorrect economic policy.

If this doesn't work, there are still resources for Scenario 3.

Whatever effects the political decisions regarding the climate issue will have in the world, greenhouse gas emissions will not exterminate the innovative human race. Our civilisation will continue its development, albeit possibly at a slower rate and under more difficult conditions.

In the worst case we will have to take a step back before the harm can be undone.

Trying to predict the future is always risky, but that's a risk I am willing to take. I claim that the increasing greenhouse effect is not a question of destiny. By a combination of innovative research work, powerful political initiatives, and human insight and technology, it will in time be reduced to a non-question.

There is an entire palette with solutions, waiting for us to paint a future in far brighter colours than the one we fear.

Finally, it is important to remember the fact that much progress has already been made in later years and that we have enormous energy resources at our disposal. My estimate is that emissions are decreasing faster than anyone has deemed possible, at the same time as energy can be produced cheaper than ever before.

Above all, let us not forget that man is an innovative and basically good creature, and that it is in our nature to create smart solutions.

Chapter 13
One Planet Is Enough

Abstract The standpoint that the human craving for consumption is despicable is extremely widespread. And that is understandable, especially when you look at the big container vessels traversing the seas in all directions, loaded with all kinds of foodstuffs and fancy gadgets. However, as I have tried to point out in this book, we cannot manage without consumption, because it is one of the main forces behind evolution. In my opinion, there is no contradiction between a good life and continued technological growth—on the contrary. Techno-evolution, in cooperation with human creativity and humankind's strongest drive—the instinct for survival—will ultimately result in a healthy planet.

An overwhelming chorus of critical voices claims that our materialistic lifestyle is having a devastating impact on nature. The apes, after all, are able to live an ecologically sustainable life, while we are not! As such, it is easy to believe that we could have been living as ecologically healthy as chimps if it hadn't been for that prehistoric ancestor of ours with the absurd idea of knocking two pieces of stone together instead of simply crushing bones.

Thus, the inventor of the stone axe really messed things up.

Today we are completely surrounded by a plethora of artificial gadgets—on our bodies, in our homes, at work, at school, in hospitals, when travelling—virtually everywhere. We are dependent on technology in all shapes and sizes. The development that started with the stone axe gave birth to *Homo sapiens*, and ultimately led to civilisation.

We may think of those early stone tools as crude artefacts rather than pieces of technology. But that would be a grave mistake. The first simple tools represented something much more potent than their mere practical functionality. They were an embryo, the first step on the road to something that would transcend human intelligence and might ultimately develop into a new independent species.

The hypotheses and messages in this book aim to answer some of the questions that I have asked myself during many years, when I have been pondering over the challenges of man and the possibilities of technology.

The standpoint that the human craving for consumption is despicable is extremely widespread. And that is understandable, especially when you look at the big container vessels traversing the seas in all directions, loaded with all kinds of foodstuffs and fancy gadgets—products that in most cases will soon be scrapped. However, as I have tried to point out in this book, we cannot manage without consumption, because it is one of the main forces behind evolution. Consumption has to be re-evaluated and regarded as a positive primordial force, especially when viewed in a longer time perspective.

Without our inherent instinct to improve, collect, create and—because of that—consume, only a handful of people would live a comfortable life today.

The question is: will our natural resources be sufficient?

My opinion is that we don't have to worry, because development is constantly advancing.

According to our experience from the last century, whenever a basic resource becomes sparse, the price goes up. That, in turn, triggers a search for alternatives or for entirely new technological solutions. The result will inevitably be innovations which are not dependent on expensive and rare raw materials. The recycling of used equipment will also become profitable.

Of course, there are some critical natural resources which we cannot live without and which are not dependent on the whims of any market—the air, the seas, the availability of food and other necessities provided by the eco systems.

In these cases, we may have to rely on political means of control—prohibitions or other activities to get the development back on the right track. Still, many natural resources are available in virtually unlimited quantities for our civilisation to use—like iron, calcium, silicon and solar radiation.

Today, we produce more food than we really need. Not even the sources of energy set a limit for our expansion. The strain on the natural resources is reduced as new and better technologies are developed. For example, fish farming on land can save sea environments. Emission-free industrial technology will improve the air we breathe. Non-toxic biocides can save the insects that are necessary for the eco system services.

Another beneficial effect of techno-evolution is that we tend to consume smaller amounts of material and energy for each produced item. Electronic devices are becoming smaller and lighter, engines more energy efficient, and light sources more energy saving. This is a rule with very few exceptions.

We are inclined to underestimate, or even forget, the value of the resources that humans have created. Nevertheless, the value of these artificial creations can sometimes surpass the value of some finite natural resources. When people are buried alive under the debris of a massive earthquake, it is the excavators they are waiting for. In situations like that, metals and petroleum are much more useful as components within telephones and ambulances than as natural resources hidden in the ground. In that respect, the footprint of progress is more important than the ecological one.

However, if everyone lived the way we do, wouldn't we need several more planets? Yes, of course, is the simple answer. But everyone will not live the way we do. It is true that we who drive cars today are increasing the level of greenhouse gases in the atmosphere. But the car drivers 30 years from now will not do that, because their cars will not emit carbon dioxide. And at that stage, we will not keep our houses warm at the expense of our planet's atmosphere.

Thirty years into the future, newspapers and books made of paper from decimated forests will be history. Then the forests will produce oxygen instead of paper pulp.

Do the rich countries consume at the expense of the poor ones?

Yes. It is an undeniable fact that the rich countries, with a mere fifth of the world's population, currently consume 80% of the earth's resources. It is also a fact that the industrial countries have been responsible for the largest emissions of greenhouse gases into the atmosphere for a long period of time.

However, the developing countries of today have one advantage: a great part of the job is already done. Therefore, their populations will reach an acceptable standard of living much faster than the currently developed nations did.

This doesn't mean that the industrialised countries should feel free to ignore their responsibility to address the global environmental problems. However, they don't need to act out of a sense of guilt, but rather because they have greater resources than the poor countries to do something about it.

What has happened in later years is that we now have a much more well-founded knowledge about climate changes, and certain political agreements in force. In addition to that, the scientific as well as the industrial communities have accelerated their efforts to provide emission-free technological solutions.

Actually, our starting position is excellent. Humankind possesses several sources of energy and each one can provide an abundance of power for the inhabitants of our planet: solar, wind and hydro power, geothermal energy as well as fusion and fission power.

In fact, our demand for energy is dwarfed by the mass of existing and potential energy sources that are available on the planet.

Today, there are several game changers which will soon appear on the stage. All this means that we actually *can* have continued growth without causing the planet's ecological systems to collapse. We just have to set our possibilities free.

In the latest 50 years, techno-evolution has dramatically reduced the carbon dioxide emissions in relation to all the work produced by the technological equipment we use.

This has happened without any international agreements. In fact, our civilisation owes its existence to techno-evolution, but the process can be disrupted by misguided politics. On the other hand, political decisions can also accelerate the technological changes in the right direction. Everything depends on how the means of control are designed and how much leeway is granted to the constructive driving forces. In my opinion, scientists and politicians have to become more aware of important drivers like techno-evolution and take them into account. Otherwise, the drawbacks of political control may well outweigh the benefits.

Technology is in a state of continuous development, and now the time is ripe for new solutions that will change society fast and irreversibly. This kind of so-called disruptive technology is so revolutionary that it will wipe out established professions, products and entire industries. Revolutionary technologies in logistics, energy production, energy storage, and power distribution are knocking on the door. In my opinion, this means that the problems with greenhouse gases will be solved even sooner than we could imagine.

I am convinced that we are now approaching a much more socially, economically and ecologically sustainable world community, as an unavoidable consequence of the evolution.

Attempts to predict the future are always risky, but I believe that the progress made during the last century will be regarded as fairly trivial in comparison to the improvements that will take place in the next 50 years. This statement is based on the fact that technology has been proved to be developing at a continuously faster pace. We must assume that scientific and technological development will continue to follow an exponential curve, as there is nothing that seems to indicate the opposite. Thus, we can also assume that further innovations will provide the main solutions for the climate and environmental problems, as well as for the remaining sickness and poverty in the world.

Still, we should not regard our progress thus far as an excuse to lean back and rest on our laurels. Rather, we should look at our achievements as a source of inspiration, and as evidence of our ability to come up with even better solutions.

Technology has always been a supporting platform for humankind's journey under harsh natural conditions to the welfare society of today. Nevertheless, we are only at the beginning of the technological era. The next phase, which we may call the second machine age, will be an accelerating journey in harmony with Mother Earth.

Humankind has always pursued a good life to the greatest extent possible. We solve problems and create improvements all the time. It is a built-in human trait, one of our basic needs, and a strong instinct. Human capital, the most important of our natural resources, is invaluable. Because of it, we are not helpless when hit by tsunamis, pandemics and agricultural disasters—instead, we face the problems and solve them using our technology and knowledge.

In my opinion, there is no contradiction between a good life and continued technological growth—on the contrary. Techno-evolution, in cooperation with human creativity and humankind's strongest drive—the instinct for survival—will ultimately result in a healthy planet.

One planet is enough.

Appendix
More Future Power to Change the Planet

There is a long list of energy sources and energy efficiency methods that have the potential capacity to change the future. In this appendix, I have collected a few of them which deserve to be mentioned, in addition to the ones that I have already discussed. There is no lack of promising solutions for the energy problem that may guide the development into the right direction:

- *Geothermal energy*—Geothermal energy technology can be divided into high temperature technology for heating and high temperature technology for electricity production. Ground source heating systems, as I and many others use for indoor heating, is a kind of geothermal energy. The energy is collected from about 200 m under the surface, even though the temperature at that level is only 4–10 °C.

  Systems for collecting heat from the ground, the lakes and the air are all examples of using a geothermal heat pump to salvage energy even from sources of relatively low temperatures.

  High temperature technology is used to extract energy in volcanic regions, or by drilling deep into the hot bowels of the earth. The purpose is usually to produce electric energy. Geothermal energy is considered to be renewable almost to the same extent as direct and indirect solar energy, as the reserves are practically limitless.

- *Natural gas*—Primarily methane gas, a fossil fuel with emissions of approximately 30% less CO_2 equivalents than petroleum.

  In comparison with coal, natural gas has about 50% less CO_2 emissions when used for energy production. Even though natural gas is a fossil fuel, we will still be needing it and similar finite sources of energy for several decades into the future. Natural gas is estimated to contribute with 40% of the global energy consumption increase between the years 2014 and 2040.

- *Fission power*—The conventional nuclear power plant may experience a renaissance with the aid of a new technology—the thorium reactor. Thorium is available in much larger volumes than uranium, and the waste is not as readily usable for nuclear weapons. Thorium—in combination with existing reactor solutions—may turn nuclear power into an important option in some countries.

There are also other possibilities for development. For example, the American energy authorities provide financial support to the development of the pebble-bed reactor, where the risk of a nuclear meltdown is considered to have been eliminated. Both India and China, with their rapidly growing economies, will invest in new nuclear power plants.

At the time of writing, there are 400 nuclear power reactors in operation in the world. In addition, there are 73 new nuclear power plants under construction, 170 are being planned and another 300 or more have been presented for consideration. This shows that nuclear power still has a part to play in a global perspective.

An advantage of nuclear power, in comparison with wind and hydro power, is that it doesn't requires much space. This makes it suited for supplying energy to huge cities with tens of millions of people.

A solar cell facility requires an area of 15 × 15 km to produce the same amount of energy as one medium-sized nuclear power plant. Electricity production in the night-time and in cloudy weather also gives nuclear power advantages.

- *Industrial waste heat*—The use of redundant heat from process industries, waste combustion, and other similar sources is nothing new. In the Nordic countries, excess energy from, for instance, the pulp and paper industry has been used for decades. But there are additional sources that can be utilised as new technology is developed, and in a global perspective, there are considerable reserves of waste heat that can potentially be salvaged. The surplus energy from large data centres is just one example. The investment cost may be an obstacle, but subventions that make the ROI time shorter can make such investments more attractive.
- *Development of conventional power plant technology*—All man-made devices, from engines and boilers to electrical equipment, are made more efficient all the time. Most of these minute improvements are made and accepted without much attention. But higher efficiency and smarter technological solutions mean reduced energy consumption and, as a result, less emissions.

The total level of efficiency has been considerably increased during a time span of 30 years. For example, the efficiency of the bigger diesel and gas engines manufactured by the Wärtsilä company is about 50% today, while it was a mere 42% 30 years ago. In other words, an improvement of 20%. When these engines are used for electricity production, the efficiency may reach 55%, and if the excess heat is salvaged, it will be no less than 90%.

- *Natural photosynthesis*—This technology makes it possible for various specialised organisms, like bacteria, algae, fungi and trees, to produce fuels. Presently, there are ongoing experiments to use, for example, microorganisms and fungi to produce hydrocarbons, so-called myco-diesel. This is yet another contribution to energy production without adding external carbon dioxide, in addition to the biological methods mentioned earlier, like anaerobic digestion of biologic matter. However, it is difficult to estimate the state of this technology 20 years from now.

Saving power

Increased efficiency and savings thanks to new technology may also dramatically reduce our energy demand. In almost all areas of our lives, there is room for improvements of our energy consumption with the aid of numerous technological inventions.

- *Lighting in a new light*   Considering that lighting accounts for approximately 20% of the world's electricity consumption, LED technology, in combination with intelligent control systems, has an important part to play. The light doesn't have to be on when no one needs it, and it doesn't have to shine at full strength when it isn't pitch black. The efficiency potential of LED technology is very high, roughly 80% when replacing light bulbs.
- *Increased industry efficiency*—Of all the electricity consumption in the world, industry is responsible for 40%, of which two thirds are used to drive electrical motors. Here frequency converters enter the picture. A frequency converter is, simply put, a counterpart to the gas pedal of a car. Many electrical motors constantly run at full speed. With a frequency converter, however, they can slow down and save energy when full power isn't needed. In industry, an increased use of frequency converters has already dramatically reduced electricity consumption, and this development continues.
- *Smart infrastructure*—Smart cities, efficient public transport, and smart electric grids can save considerable amounts of energy by optimising energy supply, traffic and lighting with the aid of modern technology. Traffic control can be optimised in various ways by mapping shorter driving routes and reducing idling. With improved overall planning, the potential for heat recovery, and district heating and cooling, is very high in cities.
- *The breakthrough of The Internet of Things, IoT*—The Internet of Things means that various technological devices can communicate with each other in a network. This phenomenon has already existed for some time on a smaller scale, for example with key cards and electronic locks being able to communicate. The new properties of IoT are that communication is possible regardless of location and that large amounts of information can be transferred, thanks to the high transmission rate.

One example is that a computer in a building can monitor the variations of the electricity price and start the indoor heating system during the hours when the price is at its lowest. Systems like that are already in operation.

Another example is intelligent speed control. It may, for example, involve information transferred to buses from sensors along the route. Data like that can adapt the speed automatically or via the driver. The criteria may be safety or fuel consumption. This kind of technology may save lots of energy and maintenance costs.

- *Smart electricity systems coordinate supply and demand*—With smart electricity systems, the electricity delivered can be controlled according to supply and actual demand. Electricity for various purposes can be bought at the times of the day when the price is at its lowest—or sold when there is a local surplus.

The automatic system keeps track of both the market price and the demand and manages buying and selling transactions without human interference. Energy cloud services are emerging in pace with the new smart technology.

- *Energy storage for cloudy days*—Energy storage, especially the storing of electricity in batteries, will also contribute to energy efficiency and reduced emissions. The demand for this technology is increasing due to the growth of wind and solar power. It would obviously be beneficial if some of the daylight solar energy could be stored for use during the night, or if wind power could be saved for calm days.

Large-scale batteries for this purpose are installed in California, where the electric energy from wind turbines are stored at a capacity of 32 MW h. Admittedly, this capacity is rather small, and the technology is in its infancy. Another method is to use surplus energy to pump up water and then let it stream back through a turbine-driven generator. Small-scale battery storage for individual consumers is on its way. This development may accelerate rapidly and become revolutionary.

- *Weight-reducing material technology*—Sophisticated materials like graphene, and nano technology are currently being developed by thousands of scientists and engineers all over the world. The results will make themselves known in many ways in the future and will probably have important effects already in the two next decades. New materials can optimise energy consumption in several ways. Above all, the weight of vehicles and moving industry components will be reduced. These materials can also reduce the friction of moving parts. Electrical distribution and storage are other areas that may benefit from this development.
- *Hydrogen gas makes clean driving possible*—Hydrogen gas produced by renewable energy sources can make completely emission-free vehicle operation possible in the future. Contrary to gasoline and diesel, hydrogen gas doesn't produce carbon dioxide or other undesired substances. If the hydrogen is produced by energy systems without greenhouse gas emissions, the hydrogen operation is entirely CO_2 neutral. The hydrogen and oxygen from the air can be converted to electricity by fuel cells, in which case the electricity is produced directly in the car. If the car is battery-powered, the electricity is extracted from the public electricity grid.

This process is beneficial to the climate and the wallet alike. And finally, we will have arrived at a point where we can drive anywhere without giving a thought to the greenhouse effect.

A new era

New behaviour patterns and system shifts can change the future in different ways:

- *Communications technology increases efficiency*—The Internet of Things, with new sensors and sophisticated algorithms, will affect most things around us.
- *A new Green Revolution is on its way*—Genetically modified livestock and crops can have a revolutionary impact, and, if wisely managed, this development may be beneficial to both agriculture and climate. It is conceivable that many crops that currently need high or low temperature, lots of water or special insecticides can be modified to more robust strains. All things considered, this means less emissions of greenhouse gases.
- *The fish are moving ashore*—Conventional fishing is increasingly being replaced by energy saving fish and shellfish farming on land. We are approaching a "blue revolution" and have oceans of possibilities at our disposal. In Khulna in Bangladesh, for example, freshwater shrimps and fish are raised in ponds, and the adjacent rice fields are fertilised by the waste and left-overs of the fish. This combination and polyculture has tripled the production with minimal environmental effects due to a simple recycling process.

This example shows that water farming doesn't have to be a large-scale venture—it can also be implemented in a small scale with modest investments. No less than 71% of the earth's surface is covered by water, but as yet, only 2% of our food comes from seas and lakes. Thus, the potential for this source of food is huge.

- *Renewable "fuels" don't need to be transported*—The impending transition from fossil fuels to alternatives will constitute a large system shift. Energy production by solar, wind and hydro power plants will have considerable advantages in comparison with petroleum, coal, biomass and gas. In addition to other benefits, these renewable fuels do not need to be transported over long distances to power plants, as the energy is produced right at the source.

In contrast, the journey oil takes from the earth's crust to the lamp is long. First the petroleum must be located, the prospecting planned, drilling facilities constructed. Then the oil must be extracted and transported, stored, refined, stored again, and finally distributed to a power plant where it is transformed into electricity so that the light can be turned on.

Solar energy can be accessed in a much simpler way. Simply put, the sun shines upon a group of solar cells which produce electricity directly. There is no need for gigantic processing facilities, long transportation and intermediate storage. In other words, we get rid of the entire supply and waste management chain that also produce emissions.

Promising solutions

There is also a plethora of promising new solutions which, if realised, may totally change the energy production in the future.

- *Artificial photosynthesis* can produce fuels according to the same principles as the photosynthesis in nature; that is, it can transform solar energy into chemical energy. The advantage of artificial photosynthesis is that the input resources—sunlight, water and carbon dioxide—are virtually infinite.
- *Geothermal deep technology* is another possible source of basic energy. As we all know, the bowels of the earth are filled with magma, a glowing, extremely hot mass that sometimes penetrates the crust in volcanic eruptions. In some places, very high temperatures can be found close to the surface, and in such locations, large power plants for the production of electricity are already in operation. With sufficiently developed drilling technology, this energy source may become available all over the world, since deeper boreholes can reach higher temperatures.

This geothermal deep-hole energy would by itself be sufficient for several world populations of energy consuming people for millions of years.

- *Large-scale production of bio oil and biogas* may replace fossil oil and gas.
- *Genetically modified organisms are turned into energy producers*—These are bacteria, fungi, plants, etc. which can be modified to produce various kinds of fuel.
- *Superconductivity* is a technology that allows for electricity distribution without energy loss. Superconductivity is already used today in applications for science and medical technology. When further developed, superconductivity may be of revolutionary importance, as it would also offer a solution for the energy storage issue. With this technology, solar energy would be readily available 24 h a day. The energy losses associated with electricity transfer and Maglev trains would be close to zero. The greatest optimists even claim that this technology alone would be sufficient to meet all the energy challenges of the world.
- *Fuel cells* are already in limited commercial use, but they will most certainly become established as an important technology in the future. The fuel cell technology is basically converting chemical energy into electricity (usually from a reaction between hydrogen and oxygen), and it has a great potential for large development leaps because the materials involved are being developed so rapidly. It is not impossible that the fuel cells will be of revolutionary importance for cars and boats, where they may replace fossil gasoline, diesel and natural gas.

The closest competitor to fuel cells is electric operation with rechargeable batteries. This development has already exceeded the expectations regarding price and storage capacity during the last ten years. One of these two solutions will probably dominate, although both have excellent opportunities to change the automotive and transport sectors entirely. My assessment is that the entire vehicle fleet will be replaced within 20–50 years, and then entirely based on carbon dioxide free operation.

Literature

af Geijerstam Jan, Industriland, 2008, Premiss
Ahlbeck Jarl, Tänk på miljön, köp ny bil!, 2001, Schildts
Amblee R.S, The art of looking into the future, 2011, Gloture
Ammenberg Jonas, Miljöteknik för en hållbar utveckling, 2014, Studentlitteratur
Andersson Martin, Gunnarsson Christer, Hållbarhetsmyten, 2011, SNS
Andréasson Per-Gunnar, Geobiosfären, 2015, Studentlitteratur
Apunyu Bonny, Mobile phones for health education, 2011, Lambert Academic Publishing
Areskoug Mats, Eliasson Per, Energi för hållbar utveckling, 2012, Studentlitteratur
Arthur W. Brian, The nature of technology, 2009, Penguin
Axelsson Svante, Vår tid är nu, 2014, Ordfront
Azar Christian, Makten över klimatet, 2009, Bonnier
Bailey Ronald, The end of doom, 2015, Thomas Dunne Books
Barrat James, Our final invention, 2015, St. Martins Griffin
Basalla George, The evolution of technology, 1988, Cambridge Unversity Press
Berkun Scott, The myths of innovation, 2010, O'Reilly & Associates
Berleant Daniel, The human race to the future, 2015, Lifeboat Foundation
Bern Lars, Antropocen, 2015, Recito
Bostrom Nick, Superintelligence: Paths, Dangers, Strategies, 2014, Oxford University Press
Brockman John, This will change everything, 2010, Harper
Brown Lester R., Flavin Sandra Postel, Världens chans, 1992, Naturskyddsföreningen, Sverige/Worldwatch Institute
Brown Lester R. m.fl, Tillståndet i världen '92, 1992, Naturskyddsföreningen, Sverige
Brox Jane, Brilliant—the evolution of artificial light, 2011, Mariner Books
Brynjolfsson Erik, McAfee Andrew, The second machine age, 2014, W.W.Norton & Co.
Brännlund Runar, Kriström B, Miljöekonomi, 2012, Studentlitteratur
Burnett Judith, Senker Peter, Walker Kathy, The myths of technology, 2008, Peter Lang Pub.
Carlson Raul, Pålsson Ann Christin, Livscykelanalys ringar på vattnet, 2011, SIS
Chang Ha-Joon, 23 Things They Don't Tell You About Capitalism, 2012, Bloomsbury Press
Charpentier Ljungqvist Fredrik, Global nedkylning, 2009, Norstedts
Christensson Erik, Lindgren Janerik, Två millennier, tvåhunda milstolpar, 1999, Ekerlids.
Chu Ted, Human Purpose and Transhuman Potential, 2014, Origin Press
Clark Andy, Natural-Born Cyborgs, 2003, Oxford University Press
Conard Edward, Unintended consequences, 2012, Penguin
Dahlen Erika, Jorden vi ärvde, 2008, Ordfront
Darwin Charles, Om arternas uppkomst, 2009, Natur & Miljö
Dawkins Richard, Den själviska genen, 1989, Prisma
Dawkins Richard, Så gick det till, 2010, Leopard förlag
Day David, The eco wars, 1989, Paladin
de Waal Frans, Vår inre apa, 2005, Svenska förlaget

Diamandis Peter H., Kotler Steven, Abundance, 2012, Freepress
Diamond Jared, Collapse, 2006, Penguin Books
Donoso Jerez Beatriz, Evolution of technological innovations, 2011, LAP Lambert Acad.
Drexler K Eric, Radical Abundance, 2013, Public Affairs
Dyer Jeff, Gregersen Hal, Christensen M, The innovators DNA, 2011, Harward
Dyson George, Darwin among the machines, 1997, Basic Books
Eklund Klas, Tillväxt, 2015, Studentlitteratur
Elert Niklas, Människoapans utmaning, 2014, Timbro
Erlich Paul, The population bomb, 1968, Sierra Club
Flavin Christopher, Tillståndet i världen 2002, 2002, Naturvårdsverket/worldwatch institute
Ford Martin, The lights in the tunnel, 2009, Acculant Publishing
Forsberg Björn, Omställningens tid, 2012, Karneval förlag
Friedman George, The next 100 years, 2009, Anchor Books
Fritz Martin, Gullers Peter, Hammarén Maria, Det industriella Sverige, 2002, Dialoger
Frostegård Johan, Den ekonomiska människans fall, 2014, Karolinska Institutet Univ.Press
Frycklund Jonas, Yppighetens nytta, 2007, Timbro
Fölster Stefan, Farväl till världsundergången, 2008, Albert Bonniers förlag
Fölster Stefan, Robotrevolutionen, 2015, Volante
Geary David, The Origin of Mind: Evolution of Brain, Cognition and General Intelligence, 2005, American Psychological Association
Gerholm Tor Ragnar, Futurum Exaktum, 1972, Alders/Bonniers och Brombergs 1999
Goldemberg José, Johansson Amulya K N, Reddy Robert H, Energy for a sustainable world, 1987, World Resources Institute
Gore Al, Earth in the Balance, 1992, Houghton Mifflin Co.
Gore Al, Uppdrag jorden, 1993, Bonnier Alba
Gore Al, Vårt val, 2009, ICA förlag
Govindarajan Vijay, Innovation, Trimble, 2012, Harvard Business Review Press
Greer John Michael, Ecotechnic Future, 2009, New Society Publishers
Greer John Michael, Long Descent, 2008, New Society Publishers
Grubler Arnulf, Nakienovi Nebojsa, Nordhaus William D, Technological Change and the Environment, 2002, Resources for the future
Gulliksson Håkan, Holmgren Ulf, Hållbar utveckling, 2011, Studentlitteratur
Gärdenfors Peter, Hur homo blev sapiens, 2000, Nya doxa
Halal William E., Technology's promise, 2008, Palgrave Macmillan
Hall Carl, Miljökapitalisterna, 2012, Liber
Heckscher Eli F., Industrialismen, KF:s Bokförlag
Helm Dieter, The carbon crunch, 2012, Yale University Press
Helpman Elhanan, Tillväxtens mysterier, 2006, SNS Förlag
Hirvilammi Tuuli, Kestävän hyvinvoinnin jäljillä, 2015, Hirvilammi, KELA
Holm Fredrik, Vad är ett milöproblem, 2013, Studentlitteratur
Holmgren David How communities can adapt to peak oil and climate change, 2009, Chelsea Green Publishing Company
Holmlund Susanne, Olin Håkan, Nano revolutionen, 2013, Santeru's förlag
Hornborg Alf, Myten om maskinen, 2010, Daidalos
Hughes Thomas P, Rescuing Promethus, 1998, Vintage
Häikiö Martti, Ylitalo Essi, Bit Bang, 2013, SKS
Isaacson Walter, The Innovators, 2014, Simon & Schuster
Jablonka Eva, Lamb Marion J, Evolution in four dimensions, 2014, MIT Press Books
Jackson Tim, Välfärd utan tillväxt, 2011, Ordfront
Jewert Jenny, Gentekniken i människors vardag, 2004, Atlantis
Jiborn Magnus, Kander Astrid, Generations målet, 2013, Dialogos
Kaku Michio, Physics of the future, 2012, Penguin Books
Kelly Kevin, What technology wants, 2010, Penguin

Kettunen Niko, Paukku Timo, Kännykkä, 2014, SKS
Klein Naomi, This change everything, 2014, Allen Lane
Klimstra Jakob, Hotakainen Markus, Smart Power Generation, Wärtsilä Oy, 2011, W.A Publish
Kurzweil Ray, Are we spiritual machines, 2000, Penguin Books
Kurzweil Ray, How to create a mind, 2012, Duckworth Overlook
Kurzweil Ray, The singularity is near, 2005, Penguin Books
Larsson Ulf, The Nobel Prize: Ideas changing the world, 2014, The Nobel Museum
Lewenhaupt Tonie, Inte bara mode, 2010, Atlantis
Lindvall Johannes, Rothstein Bo, Vägar till välstånd, 2010, SNS Förlag
Ling Rich, The mobile connection, 2004, Elsevier
Lomborg Björn, How much have global problems cost the world, 2013, Cambridge Univ. Press
Lomborg Björn, How to spend 50 Billion to Make the World a Better Place, 2006, Cambridge University Press
Lundblad Niklas, Om maskiner kunde tänka, 2014, Novellix
Lynas Mark, Sex grader, 2008, Ordfront
Lynas Mark, The god species, 2011, Fourth Estate
Mahajan Shobbit, Berättelser om alla uppfinningar från forntid till nutid, 2008, H.F. Ullmann
Malm Inge, Energi och miljö, 2003, Liber
Meadows Donella, Randers Jorgen, Limits to Growth, 2004, Chelsea Green Publishing CO.
Meadows Donella, Thinking in Systems, 2008, Chelsea Green Publishing Company
Mesoudi Alex, Cultural evolution, 2011, University of Chicago
Miller Daniel, Consumption and its consequences, 2012, Polity Press
Moberg Fredrik, Simonsen, Vad är resiliens? (Publikation), Stockholm Resilience Centre
Mumford Lewis, Teknik och civilisation, 1984, Vinga press
Möller Håkan, Lyx och mode i stormaktstidens Sverige, 2014, Atlantis
Naam Ramez, The infinite resource, 2013, University Press of N.E
Nei Masatoshi, Mutation Driven Evolution, 2013, Oxford University Press
Neumayer Eric, Weak versus strong sustainability, 2013, Edward Edgar Publishing
Norberg Johan, När människan skapade världen, 2006, Timbro
Nordhaus William E, The history og lighting suggest not/Report, 1994
Nowell April, Davidson Iain, Stone Tools and the evolution of human cognition, 2011, University Press Colorado
Nygård Henry, Bara ett ringa obehag?, 2004, Åbo akademis förlag
Nygård Henry, Energisk avfalsshantering, 2015, Stormossen OyAb
Ohmae Kenichi, The next global stage, 2005, Financial Times/Prentice Hall
Olofsson Jonas, Örestig Johan, Evolutionsteori och människans natur, 2015, Natur & Kultur
Pelling Mark, Adaptation to climate change, 2010, Routledge
Persson Christel, Persson Torsten, Hållbar utveckling, 2007, Studentlitteratur
Petroski Henry, The evolution of useful things, 2010, Knoph Doubleday Publ.
Radetzki Marian, Människan, naturresurserna och biosfären, 2010, SNS
Radetzki Marian, Råvarumarknaden, 2007, SNS
Ridley Matt, The rational optimist, 2010, Harper Torch
Roberts Alice, Evolution, människans historia, 2012, Tukan
Robinson Andrew, The story of measurment, 2007, Thames & Hudson
Rockström Johan, Klum Mattias, Big world small planet, 2015, Max Ström
Rockström Johan, Omställningen till en hållbar utveckling, 2013, Volante
Rosenberg Nathan, Birdzell Jr L E, Västvärldens väg till välstånd, 2005, SNS
Rydh Carl Johan, Lindahl Johan, Tingström, Livscykelanalys—en metod för bedömning av produkter och tjänster, 2002, Studentlitteratur
Sachs Jeffrey, End of poverty, 2004, Penguin Books
Samuelsson Kurt, Hur vår moderna industri vuxit fram, 1963, Prisma
Sandström Christian, Guld och gröna jobb (Rapport), 2012, Timbro
Sandström F.Mikael, Sätt fokus på teknisk utveckling (Rapport), 2000, Sv.Naturvårdsverket

Savory Allan, Hur man får öknen att blomstra, TED.com
Schauberger Viktor, The energy evolution, 2000, Gateway
Adaptation to climate change, Schipper E. Lisa F., Burton Ian, 2009, Earthscan
Schön Lennart, En modern svensk ekonomisk historia, 2007, SNS
Segerfeldt Fredrik, Vatten till salu, 2003, Timbro
Sherman Howard J, How society makes itself, 2015, Taylor and Francis
Smart Andrew, Beyond Zero and One, 2015, OR Books
Smick David M, The world is curved, 2008, Penguin Books
Smith Laurence, The new north, the world in 2050, 2010, Profile Books
Southwood Russell, Less walk more talk, 2009, John Wiley & Sons
Spengler Oswald, Västerlandets undergång, 2013, Atlantis
Svensson Mattias, Miljöpolitik för moderater, 2015, Fores
Taylor Timothy, The artificial ape, 2010, Palgrave Macmillan
Thompson Clive, How technology is changing our minds for the better, 2014, William Collins
Tivel David E, Evolution, 2012, Dorrance Publishing
Trotzig Gustaf, Metaller, hantverkare och arkeologi, 2014, Hemslöjdens förlag
Turney Jon, The rough guide to the future, 2010, Rough Guides
Tyler Tim, Memetics, 2011, Mersenne Publishing
Venter J. Craig, Liv i ljusets hastighet, 2013, Fri Tanke
Von Wright Georg Henrik, Myten om framsteget, 1994, Bonniers
Wade Nocholas, Before the Dawn, 2007, The Penguin Press
Walker Brian, Salt David, Resilence thinking, 2006, Islands Press
Watson Richard, Future files, 2010, Nicholas Brearley Publishing
Weisman Alan, Countdown, 2013, Little, Brown & Company
Wijkman Anders, Rockström Johan, Den stora förnekelsen, 2011, Medströms
Ziman John, Technological innovation as an evolutionary process, 2000, Cambridge U. Press
Åhlström Per, Tänk kallt, 2009, Premiss
Östberg Jacob, Kaijse Lars, Konsumtion, 2010, Liber

Sources

Affärsmagasinet Forum, 2016, Svart kisel öppnar för nordisk solrevolution
Allt om Vetenskap, 7-2013, Svenska uppfinningar räddar miljön
BBC, 2014, Dokumentary, London sewage system
Bloomberg.com
Branschorganisationen Avfall Sverige
Branschorganisationen Svenskt Kött
CITEC, http://www.citec.com
Ecorunner, http://www.ecorunner.industrialecology.se
EIA, U.S. Energy Information Administration, http://www.eia.gov
EKOWEB, http://www.ekoweb.nu/marknadsrapport
Energinyheter, 2016, http://www.energinyheter.se/ kärnkraft har ännu långt kvar till sin toppnivå.
Europaparlamentets webbplats, http://www.europarl.europa.eu-energimix
European Commission, 2016, An EU Strategy on Heating and Cooling
Exxon Mobile, 2016, The outlook for energy: A view to 2040
FAO, Food and Agriculture Organization of the UN, http://www.fao.org/key facts and finding.
Fores, 2016, CCS hjälper eller stjälper klimatet
Forskning & Framsteg 4-2013, Vill vi ha framtidens kärnkraft?
G7, 2014, CLIMATE CHANGE The New Economy
Gapminder, Novus, 2013, Project Ignorance
GEA, Global Energy Assessment
Global CCS Institute, 2015, The global status of CCS
Happy Planet Index, http://www.happyplanetindex.org
IAEA, International Atomic Energy Agency, http://www.eiea.org
IEA, International Energy Agency, http://www.iea.org/world energy outlook
Illustrerad Vetenskap/ den röda tråden
IPCC, 2014, http://www.ipcc.ch/pdf/assessment-report/ar5/syr/AR5_SYR_FINAL_SPM.pdf
IPCC, 2015, Special report on Carbon Dioxide Capture and Storage, http://www.ipcc.ch/srccs_wholereport
ISWA, International Solid Waste Association
Lazard Ltd, 2015, Levelized cost of storage analysis-v.1
LRF, Lantbrukarnas riksförbund, 2012, En trygg livsmedelsförsörjning globalt och i Sverige.
LUKE, Naturinstitutet Finland
Magasinet Teknikhistoria, 2014, Stad i ljus
McKinsey, 2013, Disruptive technologies
Millennium målen, http://www.millenniummalen.nu
MIT Technology Review, 3-2015, We can now engineer the human race
MIT, Massachusetts Institute of Technology, Carbon Capture & Sequestration Technologies.
NASA, http://www.climate.nasa.gov
National Geographic, 10-2014, En ny grön revolution

National Geographic, 5-2007, Nätet dras åt, det globala fisket är i kris
Nature, 2016, Circular Economy
NATURE, 2016, http://www.nature.com/news/google-ai-algoritm
Nature, 521-2015, Dawn of technology
Nature, vol 526, 2015, How to make most of carbon dioxide
Naturvårdsverket, 2011, Klimatomställningen och det goda livet
Naturvårdsverket, 2014, Utsläpp av växthusgaser från jordbruk
Ny Teknik, elbilar, http://www.nyteknik.se
OICA, International Organization of Motor Vehicle Manufacturers
Our World in Data, 2015, Global deaths in Conflicts since 1400—Max Roser
Our World in Data, 2015, Land use in Agriculture-Max Roser, N. Ramancutty, J.A. Foley.
Overshooting Day, http://www.overshootingday.org
Renewable energy policy network, http://www.REN21
SCB, Statistiska centralbyrån
SEI, Stockholm Environmental Institute, 2015, Rapidly falling cost of battery packs... Report.
Shell, 2013, New Lens Scenarios
Svenskt näringsliv, 2014, Global klimatnytta genom svensk konkurrenskraft
Talouselämä, 2015, Vain asunnoton elää kestävästi.
Tesla motors Inc.
Tilastokeskus Statistikcentralen Finland
Tvättmaskin.net, 2015, http://www.tvättmaskin.net/tvattmaskinens-historia
UNESCO, http://www.unesco.org
WHO, Världshälsoorganisationen, http://www.who.int/gho/mortality
William D Nordhaus, 1994, Do real output and real wage measures capture reality?
World Economic Forum, http://www.weforum.org
World Happiness Report, http://www.worldhappiness.report
WVS, World Values Survey, http://www.worldvaluessurvey.org
WWF, Världsnaturfonden, http://www.wwf.se/klimatkalkyl

PGSTL 08/11/2017